A Climate for Denial

Why do some people still
reject climate change science?

Arek Sinanian

LONGUEVILLE
MEDIA

First published 2017 for Arek Sinanian by

LONGUEVILLE
MEDIA

Longueville Media Pty Ltd
PO Box 205
Haberfield NSW 2045 Australia
www.longmedia.com.au
info@longmedia.com.au
Tel. +61 2 9362 8441

For the National Library of Australia Cataloguing-in-Publication entry see nla.gov.au

978-1-68418-878-9 (print)

An excellent primer to help us understand why people are not just being wilfully reckless in the way they view the growing impacts and risks of climate change.

This thought provoking book shines a light on a range of reasons why people don't appreciate the urgency of the changes required. It rightly highlights the role of political and cultural tribalism that has made this issue a proxy for other fights and has prevented the emergence of a more constructive risk management approach.

John Connor – CEO, The Climate Institute

Arek Sinanian has written a must-read work for our time. It is clarifying for those who acknowledge climate science but are at loss to understand and communicate with those in denial, and for those who are simply uncertain given the constant, contradicting messages.

Iain Smale – Managing Director,
Pangolin Associates Pty Ltd

Arek is a passionate and pragmatic climate change specialist who has delivered an impressive and comprehensive analysis of the reasons climate science is not well understood or accepted. This book provides compelling reasons why we need to act now – to convince governments, corporations and the community to implement effective measures to combat climate change.

Derek Low – Former head of Climate Change and
Resource Efficiency, Parsons Brinckerhoff

A fascinating and personal exploration from a leading environmental expert on how human behaviour has impacted one of the world's most wicked challenges. How we communicate climate change as experts is critical to overcoming the barriers and Arek's insights are very timely in an age of 'Fake News' and pseudo-science.

Dr Ben McNeil – Founder, Thinkable.org

This book is dedicated to my sons Taveet and Daniel Sinanian and their generation who will bear the brunt of addressing the challenges of climate change that my generation will leave behind.

I am grateful to many of my friends who encouraged and supported me in writing this book. I specially want to thank David Longfield and Winsome Byrne of Longueville Media for their editing skills and encouragement and Gary Martin for his early editorial input into the manuscript.

Arek Sinanian

Arek Sinanian is an accidental author. A migrant to Australia of Armenian heritage, he arrived in Australia when he was 14 and able to speak only more than a few sentences of English. He completed a bachelor of science (engineering) degree and a master of engineering science at the University of New South Wales.

Early in his professional life, he became passionate about energy and resource efficiency and began consulting on environmental risk management, energy efficiency and greenhouse gas abatement. He became one of the earliest practitioners in greenhouse gas management and climate change risk assessment.

In 2005, Arek was accepted into the international roster of experts for the United Nations Framework Convention on Climate Change (UNFCCC) and became the only Australian voted on the prestigious six-person Accreditation Panel of the Joint Implementation Supervisory Committee (JISC) of the UNFCCC.

This book is a result of Arek's fascination with the cognitive behavioural patterns he has observed in those who can't accept the science of climate change. When confronted with seemingly intelligent, educated people who were simply unable to accept the science, he began to question what the factors were in our accepting of information and why we tend to deny certain things against all rational thought processes. Are we all prone to denying 'facts' that simply don't fit our personal identity? Do the facts somehow threaten or challenge our beliefs?

Climate change has become one of the most polarising issues of our time. Denial of the science by some, poses questions about their reasons and about the communication of complex issues to the global community.

Could the communication of climate change science have been done better? Is it a matter for more information to be made available, or is it the way information is being presented? Are there cognitive reasons for the denial? What can we learn from climate change denial? What can we learn from the failure of the debate?

This book looks at these quesitons, and more, with the aim of making sense of climate change denial.

Contents

Introduction

Climate change[1], its causes, its effects, and whether and how we should respond globally, nationally and individually is arguably one of the most divisive and polarising issues of the past decade. And it's still going on.

There is considerable and compelling scientific evidence of the causes and long-term effects of climate change. A comprehensive study[2] published in Environmental Research Papers (Letter 8, 2013) investigated the degree of scientific consensus on climate change. It examined 12,000 peer-reviewed papers in the climate change literature and found that 97.1% of those expressing a position on anthropogenic global warming (AGW) endorsed the consensus that humans are causing global warming. Yet a significant number of people do not believe[3] climate change is happening. This is partly why it is difficult to obtain public support for global agreements to tackle climate change.

There has been sustained effort to co-ordinate and organise scientific consensus and global actions through agreements, most notably the publications of the Intergovernmental Panel on Climate Change (IPCC) and the activities of the United Nations Framework Convention on Climate Change (UNFCCC). Despite these initiatives and considerable efforts to consolidate the science on climate, as well as attempts to agree on equitable and workable actions to reduce anthropogenic[4] greenhouse gas emissions, we are nowhere near the targets needed to minimise the potential catastrophic impacts.

There are many fundamental lessons to learn from this failure to communicate the messages, and for governments to act. These lessons might also apply to the way we deal with other complex issues requiring globally coordinated action and agreements from food and water supply, refugees and mass migration, to child pornography and human trafficking.

There have been numerous studies on the behavioural and psychological aspects of climate science rejection[5] – why people just don't believe it or simply can't accept it. Some of these reasons are related and denial often involves more than one factor.

This book offers a snapshot of the reasons for the rejection through denial and tries to explain the complexities of 'selling' essential messages to the average person, and winning the political support for effective and urgent action. It is intended to provide a better understanding of the reasons for the rejection of the science by a minority in the general public, and the difficulties policy makers have, in galvanising global action.

If you know someone who is a confirmed sceptic, or who finds it difficult to accept the truth of climate change science, don't be angry at them. They are human after all, and this book will hopefully explain just how human they are. Incidentally, throughout this book, the words and concepts of denial, scepticism and non-acceptance will be used interchangeably.

When I began my research, I had no idea how much insightful material I was about to find, or how interesting and complex the challenges are for climate change communication – and for securing universal agreement for action on such a hot-button subject.

Some of the behavioural and other reasons that I have tried to explain are related and climate change denial in any person often involves more than one of the reasons. Some of them may overlap. Scepticism as a human cognitive issue is complex and a full discussion of them all would require a far more qualified and specialised expertise which I don't have. It suffices to say that we are all prone to denial. But there is a big difference between denial in our personal lives and the denial of climate change science. The difference is that one is a highly personal issue which normally only affects ourselves or at the most those immediately around us, the other affects future generations.

I am hoping that this book will broaden our conversation and perhaps identify further study in areas of our understanding human reactions to complex issues and communication which can achieve a better informed and more urgent action on climate change.

Context

I have been an environmental management consultant and an adviser in energy efficiency, climate change risk management, greenhouse gas management and sustainability for 30 years. Like many others, I have been frustrated by the lack of global agreement and action on greenhouse gas abatement mechanisms and policy. While there has been considerable global effort and some agreement (such as the Kyoto Protocol) to act, it is patchy and far too slow. But I have also been intrigued by the behavioural patterns in those who reject the science of climate change. The recurring questions for me have been: *Why do some people reject the science of climate change outright?*

Why do some people accept that climate change is taking place, but not that humans are contributing to it?

So I thought it would be interesting and perhaps insightful to analyse the response of the global community (in particular, by some leaders) to the threats of climate change. This may teach us a few lessons, and tell us how we should in future galvanise public opinion to address the range of other major global issues I have nominated. A few studies show that the profile of climate change sceptics peaks significantly in the 'white, male, and over 50☐ demographic. This group is also considerably affected by ideology (but more on that later). As this is the dominant group controlling political, financial and corporate power - and making decisions on climate change policy, it is worth looking at the reasons for the group's rejection of the science and the lack of global agreements.

But there many other questions that need to be answered. What is it about climate change that divides the community and makes it so difficult for the world to address? Has there been a gap in the communication between the climate scientists, policy makers and the general public? How does the communication to different groups in community need to change so that the number of sceptics is reduced or maybe eliminated?

I believe the scientific community could have done a much better job of explaining the basic facts behind greenhouse gases and their impact on the climate. But it has not been an easy task. And as we'll find out later in this book, it's not just about explaining the facts, quite the contrary. Discussion or explanation of climate change is complex due to the many elements of greenhouse gas emissions, the natural climatic patterns and the many facets of how the climate behaves. It's also about understanding human responses to such issues.

The Earth's climate includes the oceans and currents, wind, the biosphere, upper and lower atmospheres, solar effects, glaciers, clouds, evaporation. Add to these the complexities in determining the Earth's long-term historical average temperatures and predicting temperatures based on a large number of parameters and variables. Imagine yourself being asked to reliably measure over a long period of time, the average temperature of the Earth. How would you go about it? Where would you start? How many points would you measure the temperature of? Would you measure ocean temperatures as well as air temperatures? Would you measure the air temperatures over a number different layers above the Earth's surface? As it turns out, we have a lot of data going back a long time but it is varied in its coverage and accuracy. For example, a century ago, we didn't have the sophisticated measurement instruments and satellites we have today.

The study of climate change therefore involves the advancement and integration of our knowledge in numerous areas of science which individually are very expansive and diverse. This has contributed to the high level of misconception and misunderstanding by the general public, the media and those in positions of influence or power. It has also contributed to the difficulties faced by the scientific community in fully explaining the science and convincing the global community in a cohesive and comprehensive manner. Part of the problem is that long-term trends and un-intuitive issues such as the difference between the weather and climate, have to be understood and accepted by the average man and woman. For another example, it is very difficult for people to understand why and how a degree or two increase in the average global temperature can have the predicted impacts on the climate. It is also very difficult for most people to understand why we can have both global warming and very cold winters.

Among the numerous global challenges is devising fair and balanced actions that all nations – developed and developing – can agree to implement. Not least of the difficulties are the wide gaps between developed and developing economies – the historical responsibilities for and 'ownership' of atmospheric greenhouse gas emissions – and the enormous future costs of actions required to considerably reduce global emissions. So denial of climate science by leaders and members of the general public isn't actually the main contributor to the lack of action or slow action on climate change. In fact, the vast majority of the world's population accepts the science. So, what's going on? Why are the minority of sceptics apparently slowing down global action?

While accepting the many challenges, I have unearthed a number of findings and theories, some quite unusual, in trying to understand why some people find it difficult to accept the science of climate change.

Each of the reasons for rejection or scepticism is in some way an explanation of how humans respond to many other complex issues. And there are overlaps and linkages between these cognitive reasons and our behavioural patterns. For example, in the section titles 'the need to belong', I try to explain our tribal instincts and belonging to a Climate Sceptics' Club. In the section titled 'Ideology and politics' I try to explain the influences of ideology and political leanings on climate change denial. There is clearly a link between the need to belong and ideologically-driven behaviours. Often, ideology may steer us towards a particular group's activities. Continuing acceptance within this group, and feeding our need to belong often encourages us to take part in other activities which we may not normally have taken part in. On the other hand, our social need to belong to a group may drive us closer to a particular ideology which may then affect the way we begin to view the world. When it comes to climate change scepticism, it is very likely that a person's behaviour will involve more than one of the many reasons I will present.

The structure of the book

This book is in three parts.

Part A: a brief history and a brief explanation of the basic principles of climate change and how things are now

Part B: a summary of the reasons I have identified for the denial of climate change

Part C: a few ideas on what needs to be done in the world stage and what we all need to do to progress our current position.

A few technical terms

1 GW = 1,000 MW = 1,000,000,000 Watt

1 GWh = the power generated in one hour by a 1 GW generator

1TWh = 1,000 GWh = 1,000,000 MWh = 1,000,000,000 kWh

MtOe = million tonnes of oil equivalent

CO_2 = carbon dioxide

Mt CO_2e = million tonnes of carbon dioxide equivalent

Mt CO_2 = million tonnes of carbon dioxide

Greenhouse gases include carbon dioxide (CO_2), methane (CH_4), nitrous oxide (N_2O), fluorinated gases and water vapour.

Fluorinated gases include hydrofluorocarbons, perfluorocarbons, sulfur hexafluoride, and nitrogen trifluoride. These are synthetic, powerful greenhouse gases that are emitted from a variety of industrial processes.

Greenhouse gases vary in their Global Warming Potential (GWP) and are therefore converted to carbon dioxide equivalents (CO_2e) for consistency. For example, methane has a potential of 34 times that of carbon dioxide over 100 years, but 86 times over 20 years.

Chlorofluorocarbons (CFCs), hydrofluorocarbons (HFCs), hydrochlorofluorocarbons (HCFCs), perfluorocarbons (PFCs), and sulfur hexafluoride (SF_6) are sometimes called high-GWP gases because, for a given amount of mass, they trap substantially more heat than carbon dioxide (CO_2).

Part A

Background and History

Global agreements and actions to date

Globally, regionally and in numerous countries, the past decade has seen unprecedented action to understand, clarify and address climate change. The Appendix at the end of this book provides a summary of the main international agreements and organisations involved in coordinating our knowledge and agreements on climate change.

So what is being rejected?

Climate science and the greenhouse effect began way back in the 1820s when Joseph Fourier, a mathematician and physicist, calculated that the Earth, at its distance from the sun, should be considerably colder than it was if warmed only by the effects of solar radiation. He examined various possible sources of the additional observed heat and his conclusion was that the Earth's atmosphere might act as an insulator. This is now recognised as the first proposal of what is now known as the greenhouse effect, although he did not call it that.

His argument and evidence were further strengthened by Claude Pouillet, a French physicist in 1827 and 1838, and by the experimental observations of John Tyndall, a British physicist, in 1859. The effect was more fully quantified by Svante Arrhenius a Swedish Nobel Prize winner, in 1896. However, the term 'greenhouse' was not used to refer to this effect by any of these scientists; the term was first used by Nils Gustaf Ekholm a Swedish meteorologist, in 1901.

Since then, scientific research has expanded from the early focus on the chemistry and physics of greenhouse gases, and how they behave, to numerous other disciplines including paleoclimate, volcanism, solar radiation, oceanography, atmospheric physics, computer modelling, and the Earth's axial tilt of orbit. This has resulted in an enormous amount of scientific research and knowledge, as well as data gathering and modelling particularly since the middle of the 20th century. The overwhelming conclusion is that the greenhouse gases humans have been emitting into the atmosphere since the industrial revolution are contributing to what is known as the greenhouse gas effect and global warming. And already, we get into a problem with terminology. The term 'global warming' creates a perception that the globe is continuously and 'linearly'[6] warming. That's why people say to me:

'if there's global warming taking place, how come it's so cold today?'

This, is among the many problems faced by scientists, educators and advocates in communicating the essential issues to the general public (which will be discussed later in this book).

I will not explain all the details and the vast knowledge we have accumulated regarding greenhouse gases, global warming and climate change. There are numerous web sites, academic and governmental publications that explain this extremely well. But here's a quick summary to give a context to this book:

Earth receives energy from the sun in the form of ultraviolet, visible, and near-infrared radiation. Of the total amount of solar energy available at the top of the atmosphere, some is reflected to space by the atmosphere and clouds and some absorbed by the atmosphere and clouds. Most of the rest is absorbed at the surface of the Earth. And this has been happening from the beginning of time. But here's the difference greenhouse gases make. Because it is warm, the Earth's surface radiates at wavelengths much longer than the wavelengths that were absorbed. The long-wavelength infrared heat that is trying to escape to space is more absorbent by greenhouse gases and is reflected back towards Earth and trapped by the atmosphere, thereby warming it. In addition to the absorption of solar and thermal radiation, the atmosphere gains heat by sensible and latent[7] heat flows from the Earth's surface. The atmosphere radiates energy both upwards and downwards; the part radiated downwards is absorbed by the surface of Earth. This leads to a higher equilibrium temperature than if the atmosphere were absent.

In a nutshell, greenhouse gases, which include carbon dioxide and methane, allow certain radiation to pass through to Earth. Then some of the radiation from Earth back to space is trapped.

So basically, greenhouse gases let some radiation to pass through and act as a shield for others. By doing this, they allow the sun's radiation to pass through to our planet, but the longer wavelength radiation coming for Earth simply doesn't get sent back to the cosmos and instead gets trapped, thereby heating our planet.

What has to be said here is that climate change science has always been based on the same level of rigour, peer review and scrutiny (the *Scientific Method*) from within the scientific community as any other scientific endeavour such as chemistry, physics, applied physics, engineering, pharmacology, medical science, aeronautics, astronomy, etc. Through the Scientific Method (more on this later), the same rigor has gone into climate science as for example cancer and the design of a jet aircraft. Yet it's rare for someone to refuse to get on an aircraft because they don't trust the design engineers, or someone to say 'I don't think I have cancer, it's just a means for doctors to make money'. So what's different about climate science?

Admittedly, many of the scientific advances and the things we use and see daily are more tangible, visible and proven. For example, we can observe and personally experience medical procedures that cure diseases, and we can take a flight on an aeroplane. But it's difficult to observe greenhouse gases changing the climate. More importantly perhaps, medicine and air travel are much more personal and seemingly more immediate, more current. But climate change science is about predicted events into the future. More about that later.

So there is something quite different about climate science that perhaps requires a higher level of faith in the scientific method.

There may be relatively few people who may (and do) for example question the ethics of pharmaceutical companies and maybe even specialist physicians and surgeons, but compared with climate change, these are not such a huge globally divisive issues. So, what is it that makes climate change such a polarising and divisive issue?

Are there uncertainties in climate change science? Yes, of course there are. As for many other areas of science, there are many uncertainties, particularly regarding predictions of the impacts of global warming. Climate science is still working out the correlation between greenhouse gases in the atmosphere and global temperatures. For example, what is the likely increase in average global temperatures if greenhouse gas levels are doubled (considering all the other factors that affect the Earth's climate)? Then there's working out what impact increases in global temperatures will have on the climate. But the main areas of scientific uncertainty relating to climate change are to do with the complex short- and long-term effects of climate change on the Earth's atmosphere, its ecosystems, the oceans, and the numerous secondary and tertiary effects and feedback systems.

An example of a secondary effect of climate change is the melting of the ice caps. As more of the ice caps melt, there's less ice to reflect the radiation back to space. Another is that as the Earth warms, there's more evaporation and therefore more moisture in the atmosphere. As water vapour also acts as a greenhouse gas, it reflects heat back to the surface. This is also known as positive feedback.

On top of this uncertainty, there are other factors that affect the climate over which we have no control and which we cannot accurately predict and these include volcanos, solar flares, Earth's tilt.

Another factor over which we have no control is the world's population – expected to reach 9 billion by 2050 – and the inevitable exponential increase in the consumption of resources including energy. Ironically, technological developments prolonging and saving lives, combined with the economic model encouraging consumption we have created are significant factors in the growth of greenhouse gas emissions.

So, when people say to me:

'What about population growth?
What about solar flares?

What about volcano eruptions?'

I say:

'yes, they are all factors that directly and indirectly affect global warming and climate change, and significantly, but we can't do much about them. But we can do something about our consumption patterns, and greenhouse gas emissions.'

Basically, we know the climate is changing, and that our activities are significantly contributing to this change. But we are still refining our knowledge about how the changes will affect the Earth's ecosystems and our lives, and how we will need to adapt to these changes. And this isn't an easy task considering the uncertainties around population growth, economic development, scientific developments in alternative energy sources to fossil fuels. And that's one of the reasons that scientists have at times been vague about these predictions and this has given sceptics ammunition to question and even doubt the accuracy of our knowledge.

So, that's a quick and basic summary of the background of what is being rejected by sceptics. We'll look at some of the reasons why in Part B.

What does climate change mean?

There are a number of models that predict the short- and long-term impacts of climate change. They all give different results. That's because in predicting the impacts of global warming and climate change, numerous assumptions need to be made about the future. These assumptions include various scenarios regarding current and future population growth, economic growth and consumption patterns, consequent energy usage, and technological advances. These are all then fed into predictive models that give us future greenhouse gas concentrations in the atmosphere. The models are then adjusted using what we know about solar flares, and other natural events. Then there are many variables which involve how the climate is likely to respond to global warming, and many secondary and tertiary effects and feedbacks including melting of ice caps, ocean temperatures, etc. As for any model predicting the future, there are unavoidable uncertainties. Despite the uncertainties, the whole idea of predicting future events is to take actions to avert or minimise any negative impacts. And when it comes to climate change, there are plenty of impacts we can and should do something about.

Before we begin looking at the ramifications of global warming and its effects on the climate, let's remind ourselves of the fundamental difference between climate and weather. The difference between weather and climate is a matter of time and long-term trends. While the weather changes from day to day and one summer may be a little hotter than another, climate change will mean that over a long period, the trend will be towards hotter summers. Or, over a longer period, storms will be more frequent or severe. Weather is what conditions of the atmosphere are over a relatively short period of time.

There are a lot of components to weather. Weather includes sunshine, rain, cloud cover, winds, hail, snow, sleet, temperature, humidity, precipitation, wind, and atmospheric pressure, flooding, blizzards, ice storms, thunderstorms, steady rains from a cold front or warm front, excessive heat, heat waves and more.

When we talk about climate change, we are talking about changes in long-term averages and patterns of daily weather. If the past few summers seem hotter, then the recent climate may have changed. For instance, in various parts of the world, spring may come earlier than it did say 20 years ago.

To further complicate things, there are also longer as well as shorter-term climate variations. This so-called climate variability can be represented by changes related to El Niño, La Niña, volcanic eruptions or other changes in the Earth system. People say to me 'what about El Niño and La Niña', and I say 'yes, they change the weather'. But I add that 'these are in addition to the climate change caused by the greenhouse gas effect, and we can't do anything about El Niño and La Niña'.

An easy way to view the difference is that climate is what you expect based on long term trends or patterns, like a very hot summer, and weather is what you experience, such as a hot day with possible thunderstorms.

While some climate change and extreme weather events can be seen as positive, most will be negative. Yes, the world has experienced climate changes over millions of years – some sudden and some gradual. The sudden ones would have been catastrophic with many species becoming extinct, while the gradual ones would have required adaptation by the entire ecosystem, including humans.

It can be argued that the climate change we will be experiencing can be classified as 'sudden' in terms of geological time frames. In other words, the climate change we are experiencing is, and will continue to be relatively fast compared with prehistoric events. The main issue we have in the suddenness of climate change is that the impacts of climate change we are likely to experience will affect a huge number of people because of the enormous increase in the world's population compared with prehistoric time. Also, whereas thousands of years ago, humans could migrate to higher ground or cooler or warmer areas, our urban and settled existence makes this very difficult for billions of people. The intractable problems of recent refugee migration are a prime example. While all this poses huge challenges for us to overcome, we can still do something about it whereas in the past, we couldn't.

In a nutshell, what the world is likely to experience include:

- **More frequent and more severe or extreme weather events**

  What is a severe or extreme weather event? These are unexpected, unseasonal or unusual weather events that are at the extreme end of the historical normal distribution curve of such events. They will most likely include:

  ☼ **Heat waves** or extremely hot days that are uncharacteristic for the location and season. While every region can experience hot days, particularly in the middle of summer, the higher frequency and longevity of such hot days can have significant impact on human health, social activities, power consumption and generation systems, agriculture, plants and wildlife. Yes, there have always been extreme hot days in every region. But climate models are predicting more frequent and more severe hot days, and for longer periods. Interestingly, the occasional extremely hot day (say over 40 degrees Celsius) has been shown to have a lower impact than more prolonged hotter (say over 35 degrees Celsius) days.

☼ **Cold waves** which are the result of movements of air streams which can drop temperatures suddenly. This tends to baffle and confuse people; 'how can it be so cold when there's supposed to be global warming?' Needless to say, this also provides false arguments for the sceptics. Again, such events can have devastating impacts on human health, social activities, power consumption and generation systems, agriculture, plants and wildlife.

☼ **High winds, thunderstorms, cyclones, tornados and hurricanes**. These can cause higher seas than normal and consequently more coastal damage and erosion. They can also have increased and more devastating effects on infrastructure, transport, power lines, and telecommunication systems. Who can forget the image of a swimming pool that ended up on the beach of an Australian east coast in June, 2016? However, according to the Centre for Climate and Energy Solutions[8]:

It's unclear whether climate change will increase or decrease the number of hurricanes, but warmer ocean surface temperatures and higher sea levels are expected to intensify their impacts.

Hurricanes are subject to various climate change-related influences. Warmer sea surface temperatures could intensify tropical storms wind speeds, potentially delivering more damage if they make landfall. Based on sophisticated computer modelling, scientists expect a 2-11% increase in average maximum wind speed, with more occurrences of the most intense storms. Rainfall rates during these storms are also projected to increase by about 20%.

In addition, sea level rise is likely to make future coastal storms, including hurricanes, more damaging. Globally averaged, sea level is expected to rise by 30-120 cm during the next century, which will amplify coastal storm surge. For example, sea level rise intensified the impact of Hurricane Sandy, which caused an estimated $65 billion in damages in New York, New Jersey, and Connecticut in 2012, and much of this damage was related to coastal flooding.

The connection between climate change and hurricane frequency is less straightforward.

Globally, the number of tropical storms that form each year ranges between 70 and 110, with about 40 to 60 of these storms reaching hurricane strength. But as you'd expect, records show large year-to-year changes in the number and intensity of these storms. And in any case, the mere increase in the energy provided by global warming, simply means that there will be additional energy provided to the climatic forces.

The important issue here is that global warming will increase the energy in the weather system and this is expected to over time, significantly increase the occurrence of storms activity in one form or another and some areas of the world will experience this much more than others.

☼ **High rainfall**. Again, this refers to uncharacteristic or unusual amount or duration of rain for the region or season. Prolonged periods of high rainfall tend to create floods and sudden rises in river levels. Whereas in the past, a region may have experienced a major 'one in a 100 year' flood, perhaps now these could take place every ten or twenty years. Here's what the Centre for Climate and Energy Solutions says:

Extreme precipitation events have produced more rain and become more common since the 1950s in many regions around the world.

Scientists expect these trends to continue as the planet continues to warm. Warmer air can hold more water vapour. For each degree of warming, the air's capacity for water vapour goes up by about 7%. An atmosphere with more moisture can produce more intense precipitation events, which is exactly what has been observed, averaged over large areas of the Earth.

It is important to note that increases in heavy precipitation may not always lead to an increase in total precipitation over a season or over the year. Some climate models project a decrease in moderate rainfall, and an increase in the length of dry periods, which offsets the increased precipitation falling during heavy events. This is why it's called climate change and will affect some regions more than others. Clearly, low lying areas will be most vulnerable to floods during heavy rainfall and resulting damage to infrastructure and crops.

Apart from damage to infrastructure and homes, floods can and do considerable damage to crops and livestock.

☼ **Drought**. This refers to uncharacteristic or unusual periods of low or no rainfall. Obviously, droughts have dramatic impacts on agriculture, water supply to urban highly populated areas. Again, according to the Centre for Climate and Energy Solutions:

Global warming is predicted to increase the risk of more severe or more prolonged droughts in some regions. Even in regions that may not see changes in precipitation, warmer temperatures can increase water demands and evaporation, putting greater stress on water supplies.

Estimates of future changes in seasonal or annual precipitation in a particular location are subject to considerable uncertainty; more so than estimates of future warming. However, scientists are more confident that at the global scale, relatively wet places such as the tropics and the high latitudes will get wetter, while relatively dry places in the subtropics (where most of the world's deserts are located) will become drier.

When droughts do occur, warmer temperatures can amplify their impacts. Droughts can persist through a 'positive feedback,' where very dry soils and diminished plant cover can further suppress rainfall in an already dry area. Increased temperatures enhance evaporation from soils, making a periodic occurrence of drought worse than it would be under cooler conditions.

Climate change increases the odds of worsening drought in many regions of the world in the decades ahead.

When there are longer periods of drought and floods, there are also significant risks of higher incidents of bush fires (wildfires). This is because higher rainfall results in more vegetation (fuel) and then if this is followed by longer periods of drought means drying of the vegetation. So, once again, although there have always been periods of draught and periods of high rainfall, if this cycle is exaggerated, so is the risk of bush fires.

For all these weather events, climate change models predict higher frequency and more severe events. Most regions and, in particular, richer nations, have systems in place and much of their infrastructure have inbuilt resilience and are capable of coping with much of these increased risks of damage and devastation. However, in some areas and for some infrastructure, there will need to be considerable expenditure to strengthen and make existing infrastructure more resilient to climate change. In particular, poorer nations with minimal financial and technical capabilities will be very hard hit.

No other sector of the economy is more directly and financially affected by climate change than the insurance industry. The insurance industry continuously revises its models that predict extreme weather events and their potential impact on its customers' assets. In fact, the industry has been well ahead of any other sector in keeping abreast of climate models and predictions of extreme weather events and has accurate probabilistic models of specific events for every region of the world.

According to 742 experts and decision makers in business, academia, civil society and the public sector surveyed for the 2016 edition of the World Economic Forum's annual Global Risks Report[9], the 'failure of climate-change mitigation and adaptation' is the greatest risk facing the world in the next 10 years.

One of the fallacious arguments commonly used by sceptics is that 'from the beginning of time, the world has always had extreme weather events and we've always somehow survived them'. This is entirely true but at no other time in human history has there been more at stake in terms of the vulnerability of the world population and critical infrastructure.

- **Acidification of the oceans**

When carbon dioxide (CO_2) is absorbed by seawater, chemical reactions occur that reduce the seawater pH, carbonate ion concentration, and saturation states of biologically important calcium carbonate minerals. These chemical reactions are known as 'ocean acidification'.

Calcium carbonate minerals are the building blocks for the skeletons and shells of many marine organisms. In the parts of oceans where there's marine life, the seawater is saturated with calcium carbonate minerals. This means there are abundant building blocks for calcifying organisms to build their skeletons and shells. But continued acidification is causing many parts of the ocean to become under saturated with these minerals, and this is likely to reduce the ability of some organisms to produce and maintain their shells.

Since the start of the Industrial Revolution, the pH of surface ocean waters has fallen by 0.1 pH units. This doesn't sound much but it represents a rise in acidity of about 30%. Predictions of increased CO_2 in the atmosphere indicate the oceans will continue to absorb higher amounts of CO_2 and become even more acidic. This means that by the end of this century the surface waters of the ocean could be nearly 150% more acidic, resulting in a pH that the oceans haven't experienced for more than 20 million years.

- **Sea level rise**

Sea level has risen about 20.3 cm in the past 100 years, making coastal storms more damaging and accelerating erosion. Globally, the future rise is likely to be 30-122 cm and could be even higher if glaciers in Greenland or Antarctica melt especially quickly.

The two main mechanisms causing sea levels to rise will continue to do so as long as global warming is occurring. The first is simply through the expansion of water due to an increase in ocean temperature. The second is through the melting of the ice sheets and glaciers.

In its *Fifth Assessment Report (2013)*, the IPCC found that recent observations of global average sea level rise at a rate of 2.8 to 3.6 mm per year is consistent with the sum of contributions from observed thermal ocean expansion due to:

- Rising temperatures (0.8 to 1.4 mm per year),
- Glacier melt (0.39 to 1.13 mm per year),
- Greenland ice sheet melt (0.25 to 0.41 mm per year),
- Antarctic ice sheet melt (0.16 to 0.38 mm per year), and
- Changes to land water storage (0.26 to 0.49 mm per year).

The report also concluded that if emissions continue to keep up with the worst case IPCC scenarios, global average sea level could rise by nearly 1 m by 2100 (from a 1986–2005 baseline). If emissions follow the lowest emissions scenario, then global average sea level is projected to rise by between 0.28–0.6 m by 2100 (compared to a 1986–2005 baseline).

A storm surge is a coastal flood or tsunami-like phenomenon of rising water commonly associated with low pressure weather systems (such as tropical cyclones and strong extratropical cyclones), the severity of which is affected by the shallowness and orientation of the water body relative to storm path, and the timing of tides. Most casualties during tropical cyclones occur as the result of storm surges. The two main meteorological factors contributing to a storm surge are a long fetch of winds spiraling inward toward the storm, and a low-pressure-induced dome of water drawn up under and trailing the storm's center.

The effects of sea level rise combined with storm surge are very visible and include coastal erosion and inundation of homes and infrastructure. The impacts of sea level rise alone are enough to devastate many areas of high human population and settlement. This climate change impact alone may displace considerable number of people particularly in low lying Pacific Ocean island countries and low lying areas such as Bangladesh. Considering that many low lying parts of the world and islands that are barely a metre or two above sea level, many millions of people are vulnerable to being displaced and inundated with storms and sea level rise.

- **Threats to ecosystems**

As the climate changes, many plants and animals will be forced to shift their habitats as it becomes more difficult to thrive in their current locations. Depending on the region and the changes in precipitation, plants and animals may also move to a higher or lower rainfall area. Climate change will also change migration patterns[10] of marine life and birds with the potential spread of pests and insects. For example, certain species reproduce at times when there is plenty of food for the offspring. But what will happen if the offspring arrives too early and there's inadequate food available? These impacts will not only affect biodiversity, agriculture, aquaculture and forestry, but will also increase certain human diseases such as malaria in some parts of the world. In extreme cases, species may be unable to adapt and will disappear.

This is a highly complex area but suffice to say that plants and animals depend on each other and the entire ecosystem of everything from the earthworm to the African elephant, from insects to every plant, all depend on the climate and are directly and indirectly affected by any changes to it.

Most of the ecosystem is in at least some way adaptable and resilient. Nature has a way of coping with changes as it must. But such resilience has its limits and severe or sudden changes to climate can and will affect those parts of the ecosystem most vulnerable. One example of such vulnerability is Australia's Great Barrier Reef which supports a huge ecosystem and is showing signs of considerable stress.

What has been achieved so far?

A simplistic timeline of greenhouse gases and climate change

| 50,000 years or so ago | • Humans are going about their business, trying to survive, hunting and gathering, using fire for warmth, cooking and security. Survival is the only focus and procreation is part of it.
• The carbon footprint of the average person on Earth is miniscule and it's entirely due to the burning of wood. |

2500 years ago	• Humans now use bronze for tools and are starting to farm animals and grow crops. • The carbon footprint of the average person on Earth is still miniscule and it's mainly due to the burning of wood, charcoal, natural oils and fat. • Money and material possessions are important but mainly for survival and some for spiritual reasons.
250 years ago	• The Industrial Revolution begins and the whole world is very excited about harnessing and utilising energy provided mainly by coal. • Humans begin to use coal to generate steam and it provides power for manufacturing, transport and much more. • Life expectancy and quality of life for humans begins its exponential rise. • Humans now control their environment and technology developments begin their dominance in human existence.
190 years ago	• Greenhouse gases are identified by Jean-Baptiste Fourier (although he doesn't call them that). • In the latter half of the 19th century, humans find ways to use fossil fuels in the form of coal and crude oil. Coal is used in heating and coal gas is used for lighting, cooking and heating. The internal combustion engine is invented. These developments lead to the building of factories, the mass production of goods, and more ways of using energy. This in turn means cheaper goods and higher rates of consumption. And so begins the spiral of consumerism. • Late in the 19th century, humans develop the technology to harness the energy in coal to generate electricity, which becomes the main driver of economic, social and environmental change. • The Industrial Revolution is leading to a miraculous and sustained advancement in human existence through scientific and technological developments.

115 years ago	• Humans are now aware of the effect of greenhouse gases on our atmosphere and climate. • Humans in the developed world are busy developing new ways to use fossil fuels, with little attention to the huge amounts of greenhouse gases being emitted into the atmosphere. • Compounding the growing human hunger for energy is an explosion of world population due to technology, better medical science and food production. • Humans no longer see themselves as part of nature, but believe that they can control it and take care of it. The lack of balance between development and survival is seemingly of little concern. • Greenhouse gas emissions are invisible and their accumulation in the atmosphere has subtle and long-term affects not realised by the vast majority of the world's population.
50 years ago	• Humans are beginning to realise that the consumption of resources is far exceeding the Earth's capacity. • There is an acute awareness of the environmental damage humans are causing. Identified environmental degradation includes pollution of the soil, the oceans, rivers, the food humans consume and the air they breathe. • Researchers begin to study the impacts of greenhouse gases on Earth's climate and the findings are of grave concern. The loud cries for help from experts are deemed alarmist and exaggerated. Some are listening but to most, it's all too difficult, and action to curb climate change seems much too slow. • Ironically, harnessing energy, the very thing that helped humans survive and thrive thousands of years ago, now threatens the long-term survival of the species.

20 years ago	• The Kyoto Protocol is established in 1997 to address and reduce global emissions of greenhouse gases. It binds developed nations to emission reductions while assisting to develop greenhouse gas reduction schemes in developing countries.
2 years ago	• Despite the Kyoto Protocol and its modest achievements, global economic development (particularly by China, India, South America and South East Asia) and population growth keeps pushing more and more greenhouse gases into the atmosphere. • Many parts of Africa, Asia and South America yet to develop. • If developing nations reach the same level of affluence and consumerism as the developed world, at this rate of greenhouse gas emissions, severe climate change is inevitable and that' will affect billions of people around the world, with huge consequent economic and social impact. • 195 countries adopt the Paris Agreement in December 2015, the first legally binding deal on climate change. It provides a way forward but is it enough to achieve the levels of atmospheric emissions of greenhouse gases that we need?
Now	• Renewable energy sources are being developed at an unprecedented rate, especially in the US, China and Europe. • Scientists are concerned that much more is needed. • The decarbonisation of the global economy seems inevitable, but is it being done fast enough? Many experts don't believe so. A considerable body of scientific evidence is pointing to many environmental, social and economic effects of climate change. There will need to be an urgent effort to reduce emissions of greenhouse gases if long-term and potentially catastrophic impacts are to be avoided. • The UNFCCC (COP21) Paris Agreement sets into motion the shared realisation of the urgency of the problem posed by greenhouse gas emissions, the need for cooperative action to mitigate and adapt to climate change, and signals that the end of the fossil fuel economy is inevitable.

2050	• There is at last a much more ambitious and much more specific global agreement on significant reductions of greenhouse gas emissions. • There are plenty of alternatives to fossil fuels that are cost-effective and efficient enough to allow transition from fossil fuels. • Coal-based electricity generation plants being are no longer being planned and those in operation are being gradually decommissioned. • Significant adjustments have been made by those living in developed economies, but the transition has not resulted in the levels of sacrifice feared back in the early years of the 21st century. • Humans are still driving cars, living in air-conditioned houses and working in air-conditioned buildings, flying in airplanes, but all these things are now far more energy efficient. • Climate change has occurred but humans are now on track to avert catastrophic climate change.
Around 2150	• Data indicates that the atmospheric greenhouse gas concentrations are for the first time in a century, on the way down. • There is much optimism about the future of human habitation on Earth, at least from climate change point of view.

Renewables and greenhouse gas emissions (China, US, Texas, California)

Much has been achieved globally in the areas of renewable energy and energy efficiency. In 2015 global installed renewable energy reached almost 2 million MW, a rise of 8.7% on the previous year. The threshold of 1 million electric cars[11] on the road was also exceeded at 1.26 million. This figure highlighted significant efforts by governments and industry during the past 10 years. In 2014, only about half of today's electric car stock existed. In 2005, electric cars were still measured in hundreds.

Most developed economies such as the US, Germany, China and the European Union are heading in the right direction and decarbonising. Here are just a few examples extracted from the latest publication from the International Renewable Energy Agency[12] (IRENA).

The World

In the past decade or so, there has been unprecedented rise in the renewable energy capacity installed throughout the world. Some of this has been driven by international agreements, some by regulatory measures in certain countries, regions and states, and some by the voluntary actions of corporations and governments in response to public and economic pressures.

In 2015, total installed capacity of renewable energy was 1,964,655 MW, a rise of 8.7% on the previous year. In 2014, the installed capacity of renewable energy generated 5,294,465 GWh of electricity. That's less than 22% of the total global electricity consumption.

The US

The US has been investing heavily in renewable energy for two decades and Texas and California have led the way.

In 2015, the US's total installed capacity of renewable energy was 215,117 MW, an increase of 8.7% on the previous year. In 2014, the installed capacity of renewable energy generated 559,497 GWh of electricity.

Germany

In May 2016, for the first time clean power supplied almost all of Germany's demand on a Sunday. On a sunny and windy day, solar, wind, hydro and biomass plants were supplying about 55 GW of the 63 GW being consumed, or 87%. Power prices actually went negative for several hours, meaning commercial customers were being paid to consume electricity.

In 2015, the average renewable mix was 33%[13].

Critics have argued that because of the daily fluctuations of renewables, they can have only a niche role in supplying power to major economies. But that's looking less and less likely. Germany plans to have 100% renewable energy by 2050.

China

China is leading the world in renewable energy projects. In 2015, its total installed capacity of renewable energy was 503,796 MW. This represented a rise of 16% on the previous year and more than 25.6% of the world's total renewable energy. In 2014, the installed capacity of renewable energy generated 1,253,230 GWh of electricity. This is 23.7% of all the world's electricity generated from renewable sources.

The European Union (EU)

In 2015, Europe's total installed capacity of renewable energy was 493,296 MW, a rise of 5.1% on the previous year.

The primary production of renewable energy within the EU[14] in 2014 was 196 million tonnes of oil equivalent (Mtoe), a 25.4% share of total primary energy production from all sources. The renewable energy produced within the EU increased overall by 73.1% between 2004 and 2014.

The most important source of renewable energy in 2014 was solid biofuels and renewable waste, accounting for 63.1% of primary renewables production. Hydropower contributed 16.5% and wind (11.1%) to the renewables mix. Although their levels of production remained relatively low, there was a particularly rapid expansion in the output of wind and solar energy, the latter accounting for a 6.1% share of the renewable energy produced, while geothermal energy accounted for 3.2% of the total.

The largest producer of renewable energy in 2014 were Germany (18.4%); Italy (12.1%) and France (10.7%) followed by Spain (9.2%) and Sweden (8.5%). There were considerable differences in the renewable energy mix across the EU, largely reflecting natural endowments and climatic conditions. For example, more than four fifths of the renewable energy produced in Malta (80.3%) and around two thirds of that produced in Cyprus (66.7%) was from solar energy. By contrast, close to or more than a third of the renewable energy in the relatively mountainous countries of Sweden, Croatia, Austria and Slovenia was from hydropower. Hydropower also accounted for more than a third of the renewable energy production in the former Yugoslav Republic of Macedonia, Serbia, and Montenegro, rising to a share of almost two thirds Albania, while peaking at 90.1% of the renewables energy total in Norway. More than one fifth (22.1%) of the renewable energy production in Italy was from geothermal energy sources (where active volcanic processes exist); their share that rose to 78.7% in Iceland. The share of wind power was particularly high in Ireland (51.8%) and also accounted for close to or more than one quarter of renewable energy production in Spain, the United Kingdom and Denmark.

The yearly output of renewable energy in Malta grew at an average rate of 41.3% between 2004 and 2014, although the absolute level of output remained by far the lowest in the EU. Over this same period, annual increases averaging in excess of 10.0% were recorded for Belgium (14.2% pa), the United Kingdom (12.7%) and Ireland (11.7%), while increases below 3.0% were recorded in France, Romania, Latvia, Denmark, Sweden, Croatia and Finland.

The EU's Renewable energy directive sets a binding target of 20% final energy consumption from renewable sources by 2020. To achieve this, EU countries have committed to reaching their own national renewables targets ranging from 10% in Malta to 49% in Sweden. They are also each required to have at least 10% of their transport fuels come from renewable sources by 2020.

All EU countries have adopted national renewable energy action plans to meet their targets. These include targets for electricity, heating, cooling and transport; policy measures; the mix of technologies; and the use of co-operation mechanisms.

Renewables will continue to play a key role in helping the EU meet its energy needs beyond 2020. EU countries have agreed on a target of at least 27% of final energy consumption in the EU by 2030.

Chile, Iceland, the UK and Scotland

A number of countries are achieving exceptional outcomes in renewable energy, especially Chile, Iceland, the UK and Scotland.

Chile

Solar capacity on the central power grid has more than quadrupled to 770 MW since 2013. The country is expected to install almost 1.4 GW of solar power in 2016, up from 371 MW in 2015, according to Bloomberg New Energy Finance[15]. The country has a few infrastructural deficiency issues that will need to be addressed to overcome the surplus in some parts. But what a position to be in where there is too much renewable energy supply!

Iceland

Iceland is a pioneer in using geothermal energy for space heating. Its facilities currently generate 25% of total electricity production. During the course of the 20th century, Iceland went from what was one of Europe's poorest countries, dependent upon peat and imported coal for its energy, to a country with a high standard of living where practically all stationary energy is derived from renewable resources. In 2014, about 85% of primary energy use came from indigenous renewable resources, of which 66% was geothermal. As a result of a rapid expansion in Iceland's energy intensive industry (such as aluminium processing), the demand for electricity has increased considerably. The installed generation capacity of geothermal power plants totalled 665 MW in 2013 and the production was 5.245 GWh, or 29% of the country's total electricity production.

About 90% of households are heated with geothermal energy.

In 2014 Iceland's hydroelectric power stations had an installed capacity of 1,986 MW, and generated 72% of its electricity.

Apart from geothermal energy, 73.8% of the nation's electricity is generated by hydro power, and 0.1% from fossil fuels.

Consumption of primary geothermal energy in 2004 was 79.7 PJ, approximately 53.4% of the total national consumption of primary energy, 149.1 PJ. The corresponding share for hydro power was 17.2%, petroleum was 26.3%, and coal was 3%. Plans are underway to turn Iceland into a 100% fossil-fuel-free nation in the near future. For example, Iceland's abundant geothermal energy has enabled renewable energy initiatives, such as Carbon Recycling International's carbon dioxide to methanol fuel process.

The UK

In 2015, renewable electricity generation represented 73% of total renewable energy. Renewable electricity generation increased by 29%, to 83.6 TWh from the previous year. Of this rise, 5.5 TWh was biomass[16] which rose 42%.

Solar photovoltaic increased the most in percentage terms, from 3.5 TWh to 7.6 TWh. This resulted from higher capacity, from schemes supported by the Renewables Obligation and the Feed in Tariff[17] (FiT) scheme.

Total wind generation rose 26%, partly due to more capacity but also to higher-than-average wind speeds.

Hydro-electric generation rose 6.7% in 2015 due to higher rainfall in the main catchment areas.

Onshore wind continued to be the leading technology with a 27% share, followed by plant biomass (22%), offshore wind (21%), and solar photovoltaic (9.1%).

Scotland

Its renewable electricity capacity[18] (Q1 2016) has grown steadily over the last few years with the average annual capacity increase over 635 MW since 2007. With the current mix of renewable electricity generation capacity in Scotland total over 7.88 GW, the sector is two and a half times bigger than it was at the end of 2007. Onshore wind is the biggest single source, accounting for more than 69% of installed capacity. Hydro is 20% and solar and bioenergy provide the rest.

There is significant additional capacity in development across Scotland, with projects either in planning or already consented which now total over 13 GW. Again, capacity increases in the short term will come from onshore wind, with over 3.9 GW of capacity already consented and a further 4.1 GW in planning. There is also 183 MW of bioenergy projects at various stages of development and 181 MW of wave and tidal projects either in planning or already consented.

The good news and the bad

At the risk of swamping with more data and numbers, I think it's important to give you a general feel for the current state of global energy demand and supply, and greenhouse gas emissions. Here's a general picture:

- According to the International Energy Agency[19] (IEA), the world's total energy demand (consumption) in 2013 was 9,301 Mtoe, or approximately 390,000,000,000,000 MJ.
- Of this, coal represented about 30%, oil 31%, natural gas 21%. *In total, these fossil fuels represent more than 81% of the world's sources of energy.* And there's our main problem, because it is the burning of these fossil fuels that are generating the greenhouse gas emissions going into the atmosphere.
- The emissions from the use of these fossil fuels totalled 32,190 Mt CO_2e of which 46% was from the burning of coal, 33.6% from the burning of oil, and 19.8% from the burning of natural gas.

- In 2013 the world population was approximately 7,118 million. The average carbon dioxide emissions were 4.5 Mt CO_2e per person. In Africa, they were only 0.97 Mt CO_2e per person and in Asia only 1.54 Mt CO_2e per person. And there's another problem – the huge discrepancy between the developed and the developing world.
- By contrast, the average emissions of OECD[20] countries were 9.54 Mt CO_2e per person; China's were 6.60 Mt CO_2e per person.

Where are we now?

We are where we didn't want to be. We should never have got here and need not have come this far. We knew about climate change some time ago and took too long to start to do anything about it. And we are still not doing nearly enough. That's because where we are means that even if we were to turn off every fossil-fuelled power station, every energy using device, and stopped using our cars and aeroplanes, we will still be experiencing climate change. There's simply far too much greenhouse gases in the atmosphere and it will take a very long time for the effects of these gases to diminish.

The biggest root problem is the direct link we have established between the economy and social stability. The direct link is consumption, which drives the global economy, provides jobs and the quality of life we have become used to. Add the rapid advances in technology and mass production and the model of the world we have created is that more goods and cheaper goods are produced for us all to buy. So, the more we consume, the cheaper goods and services become and the more we consume. Then technology provides bigger, faster and better models, which we feel obligated to buy. The positive result of this spiral is that this consumption provides jobs and the resulting social and political stability. This offers opportunities for further technological developments. All this is good if only it didn't require the energy we consume to achieve it all. But if the energy we used was renewable, it wouldn't matter as much.

Everything we consume, all the goods and services we use have energy embedded in them and as we've seen, most of that energy comes from fossil fuels. And the use of fossil fuels emits greenhouse gases into our atmosphere. Each piece of paper we use, each tomato we eat and each glass of water we drink has embedded energy in it. And when it comes to goods such as cars and appliances, with aluminium, glass and steel, all require significant amounts of energy to extract, process, manufacture, transport, package and to use.

It is predicted that global temperatures will rise between 1 degree Celsius and 4 degrees Celsius in the next century. And that's really frightening stuff and we should be really worried.

We know where we are and where we need to be. The following is a simplistic picture of where we are now:

1. Since the industrial revolution which started over 200 years ago, we have been burning far too much fossil fuel.

2. Our use of fossil fuels, which is mainly driven by technology, economic growth and population rises, has increased in the past few decades. The global economy is driven by energy and more than 80% of it comes from fossil fuels.

3. The resultant rise in atmospheric greenhouse gases is causing global warming and climate change. Unfortunately, because the greenhouse gases in the atmosphere will remain there for a very long time, global warming and climate change are inevitable. Even if the whole world suddenly and miraculously stopped burning fossil fuels tomorrow, we would still experience global warming and climate change due to the levels of greenhouse gases.

4. At the present and anticipated levels of greenhouse gases in the atmosphere, we are heading to global warming of more than 2 degrees Celsius in the next few decades.

5. If we are to curb this trend and the continuous increase in global average temperatures, we need to reduce greenhouse gas emissions drastically.

At the end of the book I will outline actions that should need to be taken at the global, regional and national levels. And while we can't individually do much directly to change things on the international or regional levels, we can do much to change things locally and individually.

Part B

Reasons
for
Denial

Introduction

What follows in this Part is a summary of my research into the reasons for climate change denial. I have tried to be balanced in the selection of published papers and research documents. My purpose was to find out why people would choose to ignore the overwhelming science. I want to emphasise that the reasons for denial are psychological or social in nature and I am not trained in psychology or sociology. I am therefore aware of the limitations of what is to follow.

The need to belong

Humans are social animals and need to belong to a tribe. Belonging gives us security, a sense of stability, a sense of self and a place in society. It has been fundamental to our survival as a species. It's part of our survival instinct and it's in our DNA.

Belonging to a tribe allows exchange of ideas, lessons learnt and assistance in times of need. In modern times, thousands of years after 'leaving the cave', and thousands of years after our 'hunter and gatherer' status, although our survival doesn't depend so critically on it, we still have the instinctive need to belong to a tribe. Such is the strength of this evolutionary foundation, it still helps shape our identity and self. And we do this through many different means such as religious groups and affiliations, sporting clubs, political parties, environmental and social groups.

What does this have to do with the acceptance or denial of climate change science?

The following scenario is one that I have experienced so many times in the past few years that it motivated me to my investigations into the climate denial phenomenon, and hence this book.

This is what normally happens (yes, it still does); I'm at a private or public function and am introduced to someone. As usual, we begin by finding out what brought us to the function and the other person asks what I do. I reply:

> *'I'm an environmental management consultant, have been for over 40 years, and a specialist in climate change and energy efficiency ...'*

Almost before I finish my explanation, my listener (who is invariably Caucasian, male and over 50), interrupts to say something like

> *'I don't believe in climate change'.*

Some go beyond that and say something like

> *'I think it's a rort, a sham', (or words to that effect).*

I've learnt not to react (because there's no point) but instead ask the person:

> *'And what expertise in climate change do you have? Have you done a lot of reading on this topic?'*

Without exception, the response is:

> *'I actually don't have any expertise; I just don't believe it'*

Again, I don't react. I ask for clarification saying:

> *'What gives you this belief? is it the science you don't understand?'*

Again, I either get a shrug or just silence and raised eyebrows. I continue trying to understand this person's reasoning and ask:

> *'And if it is the science you can't accept, what would it take for you to*

be convinced?'

Invariably, I'm not given a valid response, and I'm told:

'I just don't think climate change is happening and, even if it is, I don't think humans are responsible for it'.'

Such a conversation has taken place so many times, that I am astounded by its consistency. I am just as intrigued by the passion and, at times anger with which the sceptics express their position. These responses are even more astounding when I consider that most of the people are apparently intelligent, educated and generally well read and travelled.

Here is my part explanation of it in terms of branding, identity and belonging.

As stated above, the need to belong has been vital to our survival as a species. We simply would not have survived as a species had we been loners. Today, this goes well beyond survival although still plays a part in our well-being in many practical ways

We belong to numerous groups that extend well beyond nationalism, a flag, a language or religion. These groups are too large for an individual identity so we to belong to a smaller mob with a more specific identity or purpose. So apart from larger groups such as a nation, we seek small groups because this gives us a more individual identity and helps us stand out of the masses. This instinct to belong appears equally prevalent in men and women.

But there are many other subtler and far less practical aspects about the groups we belong to. There is no survival or practical reason for belonging to a sporting group and following a particular team. These types of groups – the football team, religious group, even the type of food we eat – give us a badge. And in the case of climate change denial, it can be like a badge of honour.

I cannot find any other reason that a polite, intelligent stranger, having just met me, and knowing I have considerable understanding of climate change, would tell me 'it's a sham'. This is in modern terms a 'branding exercise' on a personal level.

Interestingly, there are several clubs one can belong to on climate change. While many other global issues such as world poverty don't allow 'clubs' of different views, climate change has a number of clubs to which one can belong. Opinions on many global issues are one side or another, it's 'black or white'. We don't normally allow or celebrate 'the middle ground'. If you asked someone whether they think poverty and starvation exist and whether they are a good thing, you won't get a half way answer such as 'I don't think they exist and in any case, a bit of starvation is normal and a good thing'. Sure, there are numerous opinions on how to tackle starvation but not on whether it exists. But climate change seemingly 'allows' a number of clubs or tribes. Here are a few of the climate change clubs that one can belong to:

Club 1:   Total acceptance of the science and the significant
            contribution of human activity to climate change,
            and acceptance that urgent action is needed.

Club 2:   Acceptance of climate science and that human activity
            is contributing to it but we can't do anything about it. It's
            too far gone and difficult. Let's hope future generations
            and new technologies can deal with it. This is a very
            influential group because its members are not labelled
            sceptics but they don't want anything done about it.

Club 3:   Acceptance of climate science and that human
            activity is contributing to it but that contribution
            is not big enough to justify any major action or
            sacrifice. Again, these members are not labelled
            sceptics but they don't want anything done.

Club 4:   Acceptance of the science of climate change but not human activity's contribution to it; – it's a normal cycle that the globe has been experiencing for millions of years. Basically, accepting that climate change is taking place but humans have nothing to do with it. Again, this group's members are not labelled sceptics but they also don't want anything done about it.

Club 5:   Non-acceptance of climate science and whether climate change is happening at all.

Club 6:   Non-acceptance of climate science and whether climate change is happening – it's a conspiracy to destabilise the developed economies.

Unfortunately, there are no studies or surveys that break down the proportions of these groups. But based on a number of surveys, I think the rough distribution would look like the graph below. The important question is why the smallest clubs have the strongest impact on policies in many countries including Australia.

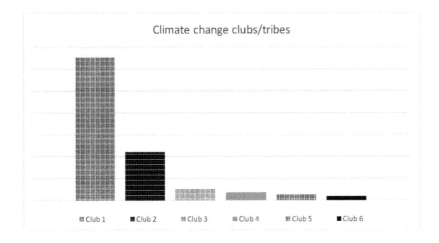

Unlike with many other global issues, belonging to one of the lesser groups can be seen as a 'badge of honour'. People may also occasionally switch from one group to another, depending on circumstance and who they are talking to.

Not least of this need to belong is that of ideology and politics which will be discussed later. Needless to say, political leaders and highly political people 'belong' much harder and need to follow the brand more strongly than others. But more about that later.

The influence of belonging to one of the climate change clubs goes further. Once we belong to one of the clubs and begin to link with other members, there is a strengthening of resolve to continue the commitment. Through social media and group exchanges, we can broaden our misinterpretations of the science (or the falsehoods) and strengthen the messages. We are also encouraged by how many others, like-minded people, also belong to this club. Just like joining a football club, we feel obligated to go to games and, once there, we are buoyed by the sheer number and enthusiasm of all our mates. Then we begin joining social networks that exchange information, reports, articles and opinion pieces. All these use confirmation bias (discussed later in the section titled 'Cherry Picking'), and strengthen our commitment to the cause. Yes, by now it has become 'a cause' and we are part of it – we truly 'belong'.

The important message out of the above is that one doesn't have to belong to any of the smaller more extreme sceptics' groups (Clubs 5 and 6) to actually disapprove or advocate for no action on climate change. Indeed, effectively, Clubs 2 to 6 all are in the sceptic camp. I'll discuss this more in the Section titled 'I don't like the solution'.

So the need to belong is an essential element for climate change denial and acceptance. The group or club we choose to belong to depends on our ideology, peers and other factors discussed below.

A little knowledge

The adage, 'a little knowledge is a dangerous thing,' is well known. Most of us know a bit about a large number of topics. Whether through formal education or through general interest and reading, we accumulate knowledge on many topics. We then use these bits of knowledge in our daily tasks and our vocation. We also make decisions and form opinions based on these bits of knowledge. But how does this play out when it comes to complex issues such as climate science?

A highly credible scientific paper by David Dunning and Justin Kruger (Cornell University, 1999) concluded that people tend to hold overly favourable views of their abilities and knowledge in many social and intellectual pursuits. This is known as the Dunning-Kruger effect. The authors suggest this occurs mostly because people who are unskilled in certain pursuits not only reach wrong conclusions and make inappropriate choices and decisions, but their incompetence prevents them realising it. The authors found that participants scoring low on humour, grammar and logic significantly overestimated their test performance and ability. For example, I might know a thing or two about making wine, and when asked to rate my knowledge of wine making, I might score myself say 8 out of ten. But in reality, my knowledge may only rate me a 5 at the most. The tests proved that people usually don't know enough about a topic to realise that they don't know enough about it to make appropriate and rational decisions. Paradoxically, improving their skills and increasing their competence helped them recognise the limitations of their abilities and knowledge. In other words, people who know very little, think they know more than they actually do; while those who know a lot, think that they know less than they actually do.

The Dunning-Kruger studies didn't test for knowledge of climate science and a sceptic's disbelief isn't a simple matter of one effect or influence. But I attribute at least some of the many reasons for disbelief of climate change science to this. When I ask sceptics how they came to their position, usually, their view is based on websites, newspaper articles, highly biased and 'cherry picked' references, or unpublished and non-reviewed articles. What is commonly known as 'cherry picking' or selectively drawing on isolated papers that challenge the consensus to the neglect of the broader body of research, is often used to argue against a number of scientific issues. This is also a factor in what is known as Illusory Correlation and Confirmation Bias explained in following sections.

There is another test I sometimes provide for those who confront me with their 'disbelief' of climate change science and it is this; I ask them:

'you're telling me that there has always been climate change and you don't believe it is any different now than it has been for millennia. What will it take for you to change your mind? What evidence do you need to prove that climate change is actually happening and that human activity is contributing to it?'

I have yet to be given a rational answer by such a sceptic.

The Dunning-Kruger effect means we often place a much higher level of reliance on our limited knowledge than we ought, particularly when it comes to complex issues such as climate change. On top of that, as we'll see later, sometimes knowledge and information on climate change matters little when it comes to accepting or 'believing' the science.

Belief and evidence

Cognitive dissonance involves the mental discomfort we all feel when conflicted between our beliefs and evidence, and when there is a contradiction or inconsistency between our actions and our beliefs.

In our complicated lives of choices, ethics, morality, the do's and don'ts, there is plenty of opportunity for cognitive dissonance. In a way, it is an inner conflict between what we should do and what we want to do. It's how we feel when we know that fatty food is bad for our health and we grab a packet of hot salty chips and eat them all.

But conflict is uncomfortable and we have many ways to overcome such conflict. To overcome or ease this discomfort and inner conflict, we look for evidence (using confirmation bias) to justify our inconsistent action. The level of discomfort depends on the level of conflict between our belief in something and the nature of action or evidence that challenges it. It also depends on the seriousness of the issue and the potential or actual consequence of our actions. To eliminate the discomfort, we may do little about it, or take extreme action, depending on the level of dissonance and discomfort.

When people are in a state of cognitive dissonance – and this is particularly true of climate change denial – they will cling even more strongly to their views, despite ever-increasing scientific evidence, to ease their discomfort. They will do this by spending more time and effort in looking for cracks in the science and any evidence they can find that creates doubt in it. They will join groups that exchange information on denial and find others with similar sceptical views. Being part of a group with similar views on the science of climate change (see the section titled *Need to Belong* above) will also ease their dissonance and discomfort.

Ironically, the more scientific evidence that is presented, the higher the level of discomfort in the sceptic. So, this will cause the sceptic to cling even more strongly to their efforts to refute or question the evidence, again to ease the higher level of discomfort. As we'll see in the next section on Confirmation Bias, this is the main reason people on either side of the fence resist being given more information that challenges their belief. They simply don't want any additional discomfort.

We all have a deep need to find consistency between our beliefs and our actions and decisions. We need to protect our belief systems and one of the effective ways to do this is through confirmation bias (see next section). Our belief system is part of who we are. It's the way we identify ourselves, our ideology and our values.

For some, global warming and climate change challenge their belief system. So, the impacts of climate change, and what we have to do and sacrifice to address it, must be either accepted or, as the sceptics choose, challenged and, if necessary, denied.

Cherry picking

The basis for prejudice

We all suffer from confirmation bias to some extent, and regularly. It is the tendency to confirm, rather than falsify, one's own belief system. This can involve a relatively trivial matter such as whether chocolate is good for our health, or a much deeper issue such as whether God exists.

One could argue that in a way, as much as I am trying to avoid it, I am doing this to some degree in this book; I'm making a case or general observations and then looking for evidence to back up my case. In my defence, when looking for evidence to explain or back up what I'm saying, I have sought out well researched, widely accepted and published information.

The pioneering research of Peter Wason in the 1960s showed we all put far more emphasis on evidence that supports our beliefs, and pay far less attention to evidence that disputes them. The result is that through a number of behavioural tendencies, we all make sure our belief is strengthened, regardless of its validity. The effect is stronger for emotionally charged issues and deeply entrenched beliefs.

Experiments Wason conducted suggested that people are biased towards confirming their existing beliefs. Later work re-interpreted these results as a tendency to test ideas in a one-sided way, focusing on one possibility and ignoring alternatives. In certain situations, this tendency can bias people's decisions and conclusions. Explanations for the observed biases include wishful thinking and the limited human capacity to process complex information.

Another explanation is that people show confirmation bias because they are weighing up the costs of being wrong, rather than investigating in a neutral, rational and scientific way. And the costs of climate change and our responses are economically, socially and morally so devastating that confirmation bias is a much easier way out.

Testing has shown that confirmation biases are prevalent at all levels of intelligence and education. Some psychologists even think that intelligent, educated people are more susceptible to it. That's because they are more likely to access and process a lot more information. They can also do research to find as much confirmation-biased information as possible. This at least partly explains the many seemingly educated, intelligent people I have met who deny climate change. And in the age of information overload technology, there's so much misinformation and so much confirmative data we can unearth, digest and share.

Even scientific investigations, research projects and academic papers can fall into the trap of using confirmation bias to prove a particular point of view or thesis. Often, a thesis is put forward and evidence is sought to prove its validity. Ideally, a thesis should explore points of view that support it and then perhaps other points of view that challenge it. The points of view that challenge the thesis should then be tested and dismissed on scientific evidence or other rational bases. Thankfully, all published scientific papers and data go through a rigorous peer review process that eliminates this (see earlier discussion on the scientific method).

A major reason for confirmation biases is related to how we encode and recall memory. Testing has shown that our memory is naturally skewed to recall memories that confirm our thoughts and feelings, especially if those thoughts and feelings are felt particularly strongly at that time.

We do this in numerous ways. One type of confirmation bias we are prone to is the process by which we give undue weight to early evidence (also known as the irrational primacy effect). By accepting early information, we risk creating a bias that affects our view of any later evidence. It seems that once our belief system is formed, it is difficult to change it. And even when early evidence is later proved to be false, we retain the original bias and our outlook is continually skewed. The climate change discourse has certainly not been helped by exaggerations and reports which, in the early days of the dialogue, were misinterpreted by self-interested bloggers and journalists. Then there's the need to belong and cognitive dissonance discussed above.

An interesting recurring experience for me is this; when confronted by a climate change sceptic who clearly has based his or her opinion on confirmation bias, I offer to explain the scientific mechanisms by saying:

'Give me 10 minutes and I'll explain the chemistry and physics of greenhouse gases, which will then give you a much better understanding of how climate change happens.'

I'm yet to be given the opportunity. This is because once we have a strong belief in something particularly emotionally charged, or one that challenges our belief system, we find it difficult to allow that belief to be shaken or challenged.

It can be argued that those like me who believe the science of climate change are just as reluctant to be persuaded otherwise and will just as passionately defend our stance.

Disconfirmation bias

There is a tendency for climate change sceptics to claim that there isn't enough evidence to support global warming and climate change. This is known as disconfirmation bias: when we are presented with evidence that disputes our beliefs, we hold that evidence to a higher scrutiny. We dispute the thoroughness or impartiality of the evidence. And we do this in a way we would never do for evidence that does not dispute our pre-conceived beliefs.

A disconfirmation bias is particularly tricky to spot because, in our own mind, it can give the impression that we are being thorough and scientific.

Here's a good example. Consider a person who through early experience or conditioning strongly believes – consciously or unconsciously – that 'older people are worse drivers than young people'. This person will accept only evidence that confirms this belief. This is a form of confirmation bias but the difference is the influence of early 'knowledge' and its strength in firming our belief. This is one of the ways prejudices are formed, perhaps through our parents, childhood friends or peers.

A disconfirmation bias is one of the main ways climate change sceptics find 'holes' and 'uncertainty' in the science. The sceptics trawl through everything to find inconsistencies and any evidence that places some doubt or uncertainty in the science. Ironically, and not surprisingly, if they spent half the time reading just some of the mountains of credible scientific papers and reports on climate change science, they would very quickly learn all about climate change. But the agenda is different to learning, and more about protecting one's beliefs and ideology (more on ideology later). Imagine someone trawling through medical records to find all the medical mistakes and wrong diagnoses to come up with the conclusion that all medicine practice is dangerous.

There is this tendency even in professions that are supposed to be rational and scientific. For example, often in studies and surveys, one starts with a premise or a thesis and then one sets out to find information that supports that premise or thesis. Usually, all effort is made to remain unbiased and balanced in the way one carries out this research or study. Nevertheless, there is an inevitable possibility of giving preference to interpretation of some of the data collected. Sometimes the researcher may simply favour (deliberately or otherwise) information that confirms the hypotheses, while giving disproportionately less consideration to alternative possibilities. But as mentioned before and later, at least the scientific method goes a long way to eliminate this through peer reviews.

People display this bias when they gather or remember information selectively, or when they interpret it in a biased way. It has been found that this effect is stronger for emotionally charged issues and for deeply entrenched beliefs. When it comes to climate change, there are considerable emotional, political, economic and other highly contentious and emotional issues (as explained in this book) that feed this bias. And to help along, these emotional issues are mixed with the many other cognitive and behavioural issues summarised in this book.

People also tend to interpret ambiguous evidence as supporting their existing position. Biased search, interpretation and memory can explain:

- attitude polarisation – when a disagreement becomes more extreme even though the parties are exposed to the same evidence

- belief perseverance – when beliefs persist after the evidence for them is shown to be false
- the irrational primacy effect (a greater reliance on information encountered early)
- illusory correlation – when people falsely perceive an association between two events or situations (explained in the next section titled 'Cause and effect').

As we can see, confirmation biases contribute to overconfidence in personal beliefs and can maintain or strengthen beliefs in the face of contrary evidence. Poor decisions due to these biases have been found in political and organisational arenas. This is not helped by exaggerations and reports that are misinterpreted by self-interested politicians, bloggers and journalists (see also later section titled 'Spreading the word').

Cause and effect

Connecting the dots
Sometimes it's difficult to connect the dots. In our daily life, we constantly make connections and correlations, establishing causes and their effects. We start doing this at a very early age to make sense of what our actions and those of others have on our world. But sometimes, we connect the wrong dots, or connect the right dots but don't necessarily connect them correctly.

Relationship perceptions or illusory correlations and their effects on our behaviour and decisions are a complex issue. Sometimes a perception can be formed that there is a relationship between events, actions and behaviours when, in fact, very weak relations exist or at times, no relationship exists at all.

Just as for confirmation bias, we all do this to some extent – mainly because it's then easier for us to make tough or complex decisions. It's a lazy way to make up our minds about a complicated issue. Relationship perceptions are mental shortcuts that usually involve focusing on one aspect or fact or event of a complex issue and ignoring others.

Imagine this hypothetical scenario: a medical specialist makes a grave error in assessing and treating someone close to you. The result turns out to be tragic and catastrophic. You've never forgotten it and never will. Rationally, you know this specialist is a highly capable, well-respected member of the medical profession. You know that these things happen and that this was a very rare and very unfortunate occurrence. But the reality is that this rare occurrence will set up a perception of failure in the medical system and a failure by this particular practitioner to save your loved one. Not only that, but it will leave you with an irrational distrust of medicine, so that from now on you will be far more sceptical and cautious about the advice you are given. And maybe this is a good thing!

Furthermore, this single highly emotional event may even take away some or much of one's trust of all medical science or science period. Mind you, such caution may actually be beneficial because it may help one to question and investigate and learn more about any advice one's given in future. It may be a very important lesson to learn and one that is part of our survival. But it may also prevent one from taking or heeding good advice in future, and this may not actually be beneficial.

Some examples of illusory correlation include:

- A child playing with a kitten receives a scratch. The next few times the child sees a kitten, it thinks that it too will scratch her.
- A man travels to a major city for the first time and is treated kindly and generously by everyone he meets. When he goes back home, he tells all his friends how kind and generous the people of that country are.
- An older person struggles to learn how to use her new smart phone. As a result, she decides not to use any other technology because it is all too difficult.

What must be said straight away is that this kind of behaviour and decision-making is reasonable and in most instances, is necessary for our social comfort and understanding. How else could a young girl protect herself from being scratched by a cat in future?

When it comes to climate change, such caution has a much bigger impact because it goes far beyond our personal circumstances. That's because one's questioning of the science affects the planet and the future of mankind. It's one thing to question science when it concerns our personal situation or problem, it's another when it concerns the whole of humanity.

Availability

Most explanations for illusory correlation involve what's known as psychological heuristics: information processing shortcuts that underlie many human judgments. Another example of these is availability: the ease with which an idea comes to mind.

Availability is often used to estimate how likely an event is or how often it occurs. This can result in illusory correlation, because some correlations can come easily and vividly to mind even though they are not especially frequent or significant. The weather is a good example because it is available and we can usually remember how cold or hot it was last week.

Working memory

Working memory is responsible for the transient holding and processing of new and already stored information. It is an important process for reasoning, comprehension, learning and memory updating.

Working memory is generally used synonymously with short-term memory but the two are different. Working memory refers to structures and processes used for temporarily storing and manipulating information. Short-term memory generally refers to the short-term storage of information and does not require the processing or organisation of information held in memory. Working memory includes the storage and processing of visual images or verbal information. Working memory tasks require processing of information or behaviours and research shows that its capacity tends to decline with age.

In an experimental study by Eder, Fiedler and Hamm-Eder[21] (2011), the effects of working-memory capacity on illusory correlations were investigated. They first looked at the individual differences in working memory, and then looked to see if that had any effect on the formation of illusory correlations. They found that individuals with higher working-memory capacity viewed minority group members more positively than individuals with lower working-memory capacity. In a second experiment, the authors looked into the effects on illusory correlations of memory load in working memory. They found that increased load in working memory led to an increase in the prevalence of illusory correlations. This meant that the development of illusory correlations was caused by deficiencies in central cognitive resources caused by the load in working memory – not by selective recall.

So, what does all this have to do with denial of climate change science? Well, it may explain how we sometimes make wrong correlations and therefore come up with erroneous conclusions, make wrong decisions, or develop (perhaps) misguided beliefs. A tendency that increases as we get older. A common one for climate change sceptics is the correlation of a cold week and therefore questioning global warming.

Religion and God

For centuries, there has been conflict between science and religion, particularly Christianity, just ask Galileo. Similarly, there appears to be a strong link between religion and God, and 'belief' in climate change science.

This may be a provocative theory but a number of studies have established such a link. A study[22] carried out in the US by the Pew Research Center concludes

Views about climate change vary by religious affiliation and level of religious observance. Hispanic Catholics (77%), like Hispanics

overall (70%) are particularly likely to say the Earth is warming due to human activity. Most of the religiously unaffiliated (64%) and 56% of black protestants say that climate change is mostly due to human activity. By comparison, fewer white mainline Protestants (41%) view climate change as primarily due to human activity. White evangelical Protestants are least likely to hold this view; 28% among this group say the Earth is warming primarily due to human activity, 33% say that the Earth's warming is mostly due to natural patterns, and 37% say there is no solid evidence that climate change is occurring.

In summary, the study indicates that while 64% of religiously unaffiliated respondents believe human activity causes the Earth's warming, only 28% of white evangelical respondents believe the same. That's an extremely large discrepancy.

In a separate study[23] by George Mason University's Center for Climate Change Communication published in March 2015, only 40% of Catholics, 39% of Protestants and 29% of evangelicals agreed that most scientists thought human-caused global warming was happening. So despite the overwhelming consensus amongst the scientific community, the perception of some Christians appears completely at odds with it.

There are a number of issues associated with the connection of God and climate change. They can be summarised by the following statements which present the possible views of some people:

1. God is omniscient and omnipotent. He controls everything and we mere humans don't have the power or capacity to change something as huge as the climate.

2. God has created the Earth and everything else for us but if we end up destroying it, so be it. That's the way it was meant to be.

3. God has created the Earth and everything else for us, and we have a responsibility to look after it.

Let's examine these statements more closely.

Humans are not capable of changing the climate

The Yale Program on Climate Change Communication presented the following statistics in a report[24] in July 2016:

> 15% of Americans think God controls the climate; therefore people can't be causing global warming

> 14% of Americans think global warming is a sign of the End of Times

Although as a species we are more powerful than any other, humans feel insignificant against the vastness of the universe. In the scheme of the cosmos, the billions of years of Earth's existence, we have a sense of being unimportant. Images sent to us from the Hubble telescope only reinforce this feeling of being inconsequential. When faced with the knowledge of hundreds of millions of years of evolution and countless calamities such as ice ages, earthquakes, floods and catastrophic volcanic eruptions, we feel powerless. Even people who have no belief in a supreme being may simply feel that they can't possibly have any impact on such a major thing as the Earth's climate.

There are so many 'forces of nature' that are far greater than what humans are capable of exerting. And for most people who have some form of **belief** in a supreme being such as God, surely, the enormity of changing the climate is beyond humans.

Associated with this feeling of actual insignificance, is the physical experiences we all have which make us have a sense of being small. We are a 'grain of sand', here on Earth but 'for a brief moment'. We have learnt that the entire presence of humans in the timeline of Earth's history is just a tiny blink in time. Even a simple trip across the world on an aeroplane gives an overwhelming sense of the enormity of the world, its oceans and skies, compared with the smallness of people.

Perhaps the best clue to the connection between religion and climate science is the often-used word 'belief'. Sceptics say, 'I don't believe climate change is caused by humans' or 'I don't believe climate change is happening'. While religion is a matter of belief and faith, science is not. Climate science is established through what is known as the scientific method with which we acquire objective knowledge based on the laws of physics and chemistry, and on empirical data, not on faith or belief. It is the same scientific method that has given us medicine, cars and aeroplanes, computers and space travel.

Even a religious agnostic may think or believe that there is a larger force, perhaps nature itself, a force much larger and much more powerful than humans, which controls climate. After all, it decides and controls the tilt of the earth, volcanic eruptions, earthquakes, etc. How often have we used the term 'force of nature' or 'an act of God' when dealing with natural calamities and events. There is even a legal term inserted in legal documents and contracts – 'force majeure', which deals with extraordinary events or circumstances beyond the control of the parties, such as hurricanes and floods.

So it is part of our normal language and general thinking that forces of nature and acts of God are beyond human intervention and climate is part of that thinking. It's similar to how some people view evolution and for similar reasons. But that's for another conversation.

Miracles can happen

Miracles are fundamental to the Christian faith, beginning with those performed by Christ – and many others attributed to the saints as it is the main criterion by which mortals can be ordained as saints.

So perhaps deep in the subconscious of Christians who deny climate change science is a belief (or hope) that God can perform a miracle to eliminate climate change.

That's the way it's meant to be

Another association of climate change with God is best summarised by a conversation I had with a man at a wedding. As soon as I explained the work I did, the man (he was Caucasian and over 50) had a simple explanation of climate change and the force of God:

'If climate change is happening and humans are going to be wiped out, then perhaps that's how God is planning it, that's how it's meant to be. Perhaps that's how human existence is meant to end, and other species will take over'.

As we were at a wedding, I didn't have the heart to ask what he would do if he was diagnosed with a life threatening but treatable disease. Would he simply accept that perhaps that's the way 'God meant it to be'? I think not.

Humans are meant to look after His creation

If the main argument of religious beliefs and climate change is that God is omniscient and omnipotent and is therefore in total control of everything including the climate, what is our role in looking after His creation? If religious climate change sceptics believe that God is in control and having created the Earth and everything else, for us humans, is God then willing to stand by and see us destroy ourselves and all the other animals and beauty surrounding us? Wouldn't God want us to preserve our species and this amazing world?

On this basis, Pope Francis released an encyclical[25] on 18 June, 2015. It was dedicated to the environment, and demanded swift 'decisive action, here and now' to head off what he sees as looming environmental ruin, and urged world leaders to hear 'the cry of the Earth and the cry of the poor'. Admittedly, the Pope's concerns were as much about global inequality, consumerism, greed and protecting the vulnerable as they were about protecting the Earth.

Another important document[26] prepared by the Lutheran World Federation in 2009 followed a global survey in which member churches identified the need to protect not only the environment as God's creation but also the most vulnerable people of the world. It said:

- In the Bible, natural occurrences such as those occurring today due to human-induced climate change, were attributed to God. People in many parts of the world still do so today. God has been considered the agent causing floods, storms, droughts and other local and global 'natural' catastrophes. People view what is occurring as being acts of God, and ask why.
- Throughout the ages, weather-related disasters have often been considered as 'acts of God'. When the destructive effects of climate change occur, some believe that God must be punishing human beings. People are told simply to wait and endure God's judgment, rather than doing anything to change what is considered to be God ordained and is inevitable.
- We maintain that somehow God is involved in climate change – specially to wake us up to the urgency of what is occurring – but we cannot attribute climate change to 'acts of God'. We must also turn to science, through which we learn more deeply, and with greater awe, about what God has created.

I don't like the solution

Even if we accept the science of climate change, we may not like what we have to do about it. This is solution aversion. And there's evidence that many who accept climate change science are reluctant to support any direct action, particularly if it threatens their quality of life or financial status.

A recent study[27] from Duke University finds that people evaluate scientific evidence based on whether they view its policy implications as desirable. If they don't, then they tend to deny the problem even exists. The researchers concluded that

the cure can be more immediately threatening than the problem.

It's like a smoker accepting the risks of respiratory disease but not accepting the consequence of putting on weight if he chooses to stop smoking. So, he denies that the risk of respiratory disease is significant or that it doesn't apply to him, or that he'll deal with the risk, or at least ignores it.

The Duke University researchers conducted three experiments on three issues: climate change, air pollution that harms lungs, and crime. The goal was to test, in a scientifically controlled manner, the question: does the desirability of a solution affect beliefs in the existence of the associated problem?

They found it does and that it occurs in response to some of the most common solutions for popularly discussed problems. On climate change, the researchers examined why some people seem to deny its existence, despite strong scientific evidence that supports it. One explanation, they found, might have more to do with opposition to the most popular solution, increasing government regulation, than with fear of the climate change problem itself, as some have suggested.

Participants in the experiment read a statement asserting that global temperatures would rise 3.2 degrees Celsius this century. They were then asked to evaluate a proposed solution to address the warming. When the solution proposed was a tax on carbon emissions or some other form of government regulation, which right-wing ideology generally opposes, only 22% of conservatives said they believed the temperatures would rise the 3.2 degrees Celsius. But when the solution proposed was free market measures such as green technology, 55% of right-wing participants agreed with the statement.

For left-wing participants, the same experiment recorded no difference in their belief, regardless of the proposed solution to climate change.

'Recognizing this effect is helpful because it allows researchers to predict not just what problems people will deny, but who will likely deny each problem'

'The more threatening a solution is to a person, the more likely that person is to deny the problem.'

The researchers found right wing-leaning individuals showed a similar aversion to solutions they saw as politically undesirable in an experiment involving violent home break-ins. When the proposed solution called for looser versus tighter gun-control laws, those with more liberal gun-control ideologies were more likely to play down the frequency of violent home break-ins. The researchers concluded:

'We should not just view some people or group as anti-science, anti-fact or hyper-scared of any problems...Instead, we should understand that certain problems have particular solutions that threaten some people and groups more than others. When we realize this, we understand those who deny the problem more and we improve our ability to better communicate with them.'

The researchers added that solution aversion can also help explain why political divides become so intractable:

'We argue that the political divide over many issues is just that, it's political...

These divides are not explained by just one party being more anti-science, but the fact that in general people deny facts that threaten their ideologies, left, right or centre.'

So basically solution aversion means that our belief in climate change may well be affected by the means of abating it, rather than the robustness of the science. Tragically, this tendency is also prevalent in other difficult social circumstances, such as personal relationships, where people often deny that there are troubles that should really be resolved.

People use all sorts of excuses for solution aversion regarding climate change. A common one is 'why should we do anything about it and make sacrifices, when the real problem is China?' If I were to give a simple answer to this question, I would say that firstly, China is investing more in renewable energy than any other nation. Secondly, that it has become a high emitter partly because we have delegated most of our manufacturing to it. Finally, that China is still catching up to our level of affluence and consumption and is still one of the lowest per capita emitters of greenhouse gases.

The same is often said about the USA in an attempt to side-step the solution and deny climate change. Once again, it turns out that – somewhat unexpectedly for some – the USA has been and is still installing large numbers of major renewable energy plants and systems and is the second highest investor in renewables. Admittedly, they have a long way to go.

There is another response to those who deny climate change through solution aversion – and it is philosophical. If there's a major global problem and we know the solutions to address it, is it a responsible response to say 'we are not the main problem here – those who are contributing more to it should fix it'? The issue becomes one of ownership and where to draw the line. Should the solution be split according to those that have contributed most to a global problem? How would you split the problem amongst all the nations of the world and how much each has historically contributed, or is contributing to, the problem? If anything, by taking positive action, even relatively small players, such as Australia[28], can put pressure on those that have proportionately more responsibility.

The world is currently having similar philosophical problems dealing with refugees. How does the world distribute humanitarian aid? By financial or economic capacity to deal with it? By the capacity to deal with ethnic or social harmony posed by emigrants?

Solution aversion is a major force in addressing climate change. A number of studies have shown that even those who accept the science may be reluctant to accept the solution.

Distrust of authority

Over the past few decades and, increasingly, as more information becomes available on just about every topic, the global community has been inundated with widely varying, usually conflicting, and often misleading 'facts'. As a result, we have all become more confused, more cautious and, in a sense, more questioning of the 'facts' presented to us. Sometimes we are perhaps even distrustful of what we are presented with and of science generally. As the world becomes more educated, more people are in a more knowledgeable and better informed position to interrogate and seek clarification, ask questions and seek answers. There is always some uncertainty in most areas of science so there is a tendency to question or seek clarification on matters relating to scientific findings and conclusions. This is a good thing.

Scientific uncertainty is the exception and not the rule but because it is usually the exceptions that get reported, there has been erosion of our unconditional trust in science.

Once upon a time, we relied on experts, academics, teachers, and those with specialist knowledge to explain the world around us, and tell us 'the facts of life'. We made our decisions or acted on these facts of life and relied on the advice of experts in their field of endeavour. Now, thanks to the internet and search engines, and to some degree at least, we can search for answers and clarifications ourselves. But as the section explained above on the Dunning-Kruger effect, this can be misleading, and even dangerous. Not only can we be selective in the answers we find, but we may also genuinely be misled by erroneous data or 'facts' or 'fake news'.

There is also the added misconception that scientists, like any other professional group with financial pressures and aspirations, are vulnerable to financial and self-interest and therefore bias. Sceptics have even suggested that some climate change scientists have gained or are likely to gain from their assertions.

Associated with a distrust of science are conspiracy theories. When the overwhelming body of scientific opinion establishes that something is true, the sceptics won't admit that scientists have reached the same conclusion by independently studying the evidence. Instead, they claim scientists are engaged in a complex and secretive conspiracy. One assertion is that the conspiracy is a communist plot to destabilise Western economies. This is similar to accusations of medical practitioners being influenced by the pharmaceutical industry. The recent movement against vaccination is yet another example and it has put childhood health at great risk.

Although not directly related to our distrust of science, or at least a lack of full trust of it, there are many other areas of our lives that we have become less trustful. These include governments, the police, clergy and religious institutions, even the judiciary. The media is full of daily reports of people in power, people in positions of trust being caught out for unethical, illegal or simply wrongful acts.

The catastrophic projections and seemingly alarmist views on climate change also complicate the messages being conveyed to the general public. In fact, many people and reporters in the sceptics' club, argue that the projections are alarmist and exaggerated, and therefore unrealistic. It is only natural that if and when a projection is made that is so alarming and disturbing, it will be viewed to be exaggerated. This is particularly the case for long-term projections of climate change which are clearly not yet experienced in full. The predicament climate change scientists face is this: let's say their findings indicate a catastrophic outcome. If they report those predictions, they will be seen as scaremongering. On the other hand, if they deliberately play down the findings (to avoid accusations of exaggeration), the public is likely to dismiss the findings as marginal or not urgent enough for radical and expensive action.

Another factor associated with distrust of authority and hierarchical organisations in our society is what is known as anti-authoritarianism. This involves general opposition to any form of authority. The main conflict here is between obedience or subjection to authority, individual freedom and determination. Anti-authoritarians do not have to be extreme in the form of anarchism, an ideology which entails completely opposing authority or hierarchical organization in the conduct of human relations. It's simply 'don't tell us what to do' or 'don't tell us what's right for us'.

The real challenge in communicating the required messages is faced when one needs to make announcements or provide existing well-established, or new details of scientific knowledge on climate change. Consider starting a sentence with:

'The science on climate change is telling us that …'

Would that work? I don't think so, because as explained above, there is already considerable distrust or at least some cynicism towards science. We could say:

'Scientists are now telling us that …'

Would that work? Unlikely. Those who have some doubts as to the reliability of scientific knowledge have an even higher level of distrust on scientists usually because it is believed that scientists as individuals, have or can have vested interests, such as vying for funds for research. So, how else could be begin such a sentence? How about:

'Experts are telling us that …'?

Distrust of science and authority in general is a complex societal issue. A big contributor to our increasing tendency to distrust or to reject what scientists or experts or authorities tell us is the constant media reporting of exceptions, frauds and failures of the 'system'. Every day, there is a barrage of news of failures and mistakes made by authorities and experts. The financial system led the world down into the Global Financial Crisis; religious groups found to have abused their trust and molested or abused young children; politicians throughout the world have been found to have abused their of power; banking executives have been found guilty of fraud; hospitals covering up diagnostic errors and surgical procedures that have gone wrong. We don't have news that says:

'This year, surgical operations throughout Australia have saved the lives of tens of thousands of sick people.'

Instead, we hear:

'This afternoon, a woman was misdiagnosed for appendicitis in a local

hospital and she was sent home to die'.

As if that's not enough, the next day, there's a follow up comment or opinion by an 'expert' who says that this event is:

'Not an isolated case'

This leads to a tendency for us to listen to experts or authorities, consider their reasoning, empirical justifications and conclusions, and then make up our minds as to whether we accept their information, judgements or conclusions. Depending on the complexity of the issue, we might listen to a number of experts or authorities and then make up our own mind as to which of them we accept or believe. An interesting aspect is how we choose which of experts' views we accept. As explained in the earlier section on *Confirmation Bias*, our choice isn't always based on reason or logic and can – ironically – be biased. I say it's ironic because one of the reasons we often reject an expert's opinion or conclusion is that it is biased. We have our own bias on what we consider to be biased.

The appeal to authority relies on an argument of the form:

'A' is an authority on a topic

'A' says something about that topic

'A' is probably correct

I'm not suggesting that authorities and experts are always right and that we should always agree and follow their advice. But if they are a true authority, then should we not listen to them, and rationally consider what they say? And for climate change, should we not study the experts' peer-reviewed reports first before forming an opinion on it? But how many of us actually do this?

We might go to our doctor to get advice on our illness and choose which part of the doctor's advice we heed. We might, for example, take the medication prescribed, but choose not to go ahead with surgery, even though we are told that surgical intervention will provide the best outcome and long term solution. Alternatively, we might take the prescribed medication, but not as often, perhaps to alleviate the side effects. Ideally, we should base such decisions on expert advice.

There is yet another social phenomenon taking place around the world with significant political and social implications. The 'Trump factor' and the recent Brexit vote in Britain are good examples. It's a combination of anti-globalisation mixed with anti-establishment and anti-politics, and community concern about local control and ownership. But just like the climate change debate, this is being argued on perceptions instead of facts. The debates have been mainly on fear and failings rather than opportunities and success.

Like other complex policy issues such as immigration and asylum seekers, climate change is seen by some as an example of political correctness gone wrong. Why else would political leanings have such a profound effect on climate change acceptance or denial? Some also see climate change as an issue raised by the elite – the political elite or the scientific elite; those who see themselves and their views above the interests of the general population. But perhaps it goes much wider than that when a simple solution is provided to a complex issue. This was the case during the Brexit debate where the 'Leave' campaign provided a deceptively simple slogan of 'Vote Leave – Take Control'. So climate change may appear to some as global action for the 'greater good' but for the loss of control for some. This will also be discussed next.

Globalisation

Globalisation is arguably one of the most important advances in the way we live. It allows the integration of the world technologically, economically and politically. Those who favour it say it has considerable benefits in:

- technological advances and sharing of knowledge and expertise, encouraging innovation
- opening up of markets, which allows the economic development of regions which might not have previously had access to them
- increased availability of goods and services; international trade in manufactured goods rose 100 fold between 1955 and 2005
- Lower costs for essential goods and services, allowing the social and economic development of many previously underdeveloped regions and countries
- employment opportunities and social stability due to easier global exchange of capabilities and skills.

But globalisation has not benefited everyone. Many communities feel they no longer control some aspects of their lives. They feel they are sacrificing too much for the greater good. Part of this is due to the recent migration of refugees and the global financial crisis. The anti-globalisation sentiment around the world has also led to movements such as Occupy Wall Street and the intensity of the debate has staged numerous protests at meetings of the World Trade Organisation (WTO), the European Union (EU), the World Bank, the International Monetary Fund (IMF), the World Economic Forum (WEF), the G8 and G20.

Those who oppose globalisation say it has led to considerable negative changes including:

- the economic dominance of powers such as the USA, the European Union and China. Although poorer nations benefit in terms of trade and job opportunities, globalisation has done little to close the gap between the very poor and the very rich
- the economic dominance of multinational corporations has been blamed for exploitation and economic opportunism in cheap labour and poor working conditions, such as 'sweat shops' in under-developed countries
- the drain of skilled people from poor countries to rich

- the introduction of fast food and poor diets to poor countries although the consequent reduction of health standards is compensated for by better healthcare
- economic vulnerability created by linking economies; during the global financial crisis systemic problems in a few regions led to a global collapse of financial markets
- the treatment of jobs in developed countries where local jobs may be lost to lower cost labour from underdeveloped countries and this affected many
- the loss of cultural and social values in developing countries due to the dominance of developed countries
- environmental degradation, particularly deforestation and land clearing, which leads to the removal of habitat of many endangered animals
- easier spread of diseases
- easier trade of drugs and illicit goods.

It's instructive to look at who voted for Brexit. There were striking and clear trends and patterns. In a crude and simplified analysis, there were correlations between:

- education level and the vote to leave; the less educated were more likely to want to leave
- salary and the vote to leave; those with lower salaries were more likely to want to leave
- age and the vote to leave; older people were more likely to want to leave
- marital status and the vote to leave; married people were more likely to want to leave
- ethnic origin and the vote to leave; people with foreign heritage or parentage were more likely to want to leave.

Correlation does not imply causation but the above shows that those who feel more vulnerable – or perceive that they are negatively affected – were more likely to want to get out of the EU. They wanted to regain control and ownership of local issues; economic, social and political.

Perceptions of risk

There are plenty more reasons, excuses and psychological explanations for denial of climate change science. Another is to do with a concept known as cultural cognition.

It appears that contrary to intuition, an understanding of science and being educated and logical do not necessarily preclude people from rejecting climate science. Of course the vast majority of logical, educated and scientifically trained people who actually understand how climate science has been established and what climate change means, know that it's real and fully accept how it works and how human activity must be contributing to it.

But, studies into cultural cognition indicate that individuals go through a variety of psychological processes to form beliefs about risky activities that match their cultural evaluations of them. People who have fairly individualistic values, for example, tend to value commerce and industry and are inclined to disbelieve that such activities pose serious environmental risks. In contrast, those who have fairly egalitarian and communitarian values, readily accept claims of environmental risks. This is consistent with their moral suspicion of commerce and industry as sources of inequality and symbols of excessive self-seeking. This is related to the link between ideology and the acceptance of climate change science but in cultural cognition, the issue relates to risks and, in particular, risks to human existence and economic growth. This also seems to be a similar factor in scepticism to *Confirmation Bias*. But this is slightly different and deals with the issue of how we perceive and deal with risk.

Scholars have provided two types of evidence to support the cultural cognition hypothesis. The first is general survey data that suggests individuals' values more strongly predict their perceptions of risk than do their race, gender, economic status or political orientations.

The second type is experiments that identify psychological processes that connect individuals' values to their beliefs about risk. These experiments suggest that individuals selectively accept or dismiss information that reinforces beliefs in line with their values. They show that individuals tend to be more persuaded by experts perceived to hold values similar to their own, rather than by those perceived to hold values different from theirs. The experiments suggest that such processes often result in divisive forms of cultural conflict over facts, but can also be managed in fashions that reduce such disagreement. In an experimental study, researchers found that subjects were substantially more likely to count a scientist (of elite credentials) as an 'expert' in his/her field of study when the scientist was apparently taking a position consistent with the one associated with the subjects' cultural predispositions. Whereas if the scientist took a contrary cultural position, his or her expertise counted for far less.

Another survey showed that members of opposing cultural groups held highly divergent impressions of what most scientific experts believed about various matters – a finding consistent with culturally biased recognition of who counts as an expert. Across a range of diverse risks (including climate change, nuclear waste disposal, and private handgun possession), the study found that members of no particular cultural group were more likely than any other to hold perceptions of scientific consensus that consistently matched those adopted in 'expert consensus reports' issued by the US National Academy of Sciences.

A study by Cultural Cognition Project researchers (using a nationally representative US sample) found that ordinary members of the public did not become more concerned about climate change as their understanding of science increased. Instead, the polarisation among cultural groups with opposing predispositions increased.

Apathy about climate change is often attributed to a lack of comprehension. It is assumed those who do not accept the science do not understand the evidence or avoid being misled. Widespread limits on technical reasoning aggravate the problem by forcing people to use unreliable cognitive heuristics to assess risk. A study by Kahan, Peters, Wittlin, Slovic, Larrimore Ouelette, Braman & Mandel[29] (Yale Law School) tested this account and found no support for it. Members of the public with the highest levels of science literacy and technical reasoning were not the most concerned about climate change. Rather, they were the ones among whom cultural polarisation was greatest. This suggests that divisions over climate change do not stem from a lack of understanding of the science. Instead they derive from a conflict between the *personal* interest individuals have in holding the same beliefs as their peers and their *collective* interest in using the best available science to promote common welfare.

Cultural cognition has been subjected to criticisms from a variety of sources. Some economists and psychologists have suggested that the theory (and those based on the cultural theory of risk generally) explain only a fraction of the variation in popular risk perceptions.

It's complicated, really

There are many complications when dealing with climate change, both by the public and by policy makers and authorities. Not least of these is the complexity of climate change science and the many disciplines involved, combined with the pace at which change is taking place. Change in our knowledge and observations, political views, economic and social impacts, and the responses by the public in general all further complicate the climate change debate.

I mentioned early in the book that the Earth's climate is so complex and so variable that it cannot be easily and reliably stated and understood. The communication about climate change science is complex – not only because of the psychological issues discussed but because of the many factors involved in the science and the way climate change works. Added to this is the subtlety of climate change. While other discrete natural events such as earthquakes, severe storms and volcanic eruptions are visible, discrete, immediate and sometimes catastrophic, climate change takes place over a long time and is difficult to observe and correlate. On the contrary, on an average day, one wakes up and looks out the window and sees the sky, the sun rising, it may be a little cold, or a little warm, or it may be raining, but everything appears normal.

There's added confusion from the non-linearity of climate change. The climate does not and will not change constantly or in a linear fashion. So the many changes in climate, will not grow constantly over time – or at an equal rate all over the world. Some parts may become cooler for a while, then warmer. The number of violent storms will not increase by one extra every year. Temperatures will not change by a fraction of a degree each year. It is the overall, long-term trends that will change.

Part of the complexity and consequent misunderstanding of the basic mechanisms of climate change relates to the difference – and confusion – between climate and weather.

Unfortunately, most people reasonably associate daily weather with climate. The most common such association is for someone to tell me

'if the Earth is warming, how come it was freezing last week?'

Or just as commonly

'if global warming is real, why did Europe and the USA have a huge freeze this winter?'

These are reasonable questions, but assume that with global warming, every point on the Earth will be hotter every day, and every season. They also assume that with global warming, there won't be any cold winters or times when some places will experience strange and yes, even colder weather.

The key to the difference is the term 'extreme weather'. Basically, global warming will increase the likelihood of extreme weather. There will be hotter days, colder days, more severe and more frequent storms. This is mainly because the higher levels of heat trapped within the Earth's atmosphere will affect wind patterns, ocean currents and atmospheric pressure systems. The trapped heat is trapped energy. The more energy trapped in our atmosphere means there's more energy for winds (a form of energy), ocean currents (another form of energy) and of course thunderstorms, which are also driven by energy. A common analogy for this trapped energy is that of a hyperactive child being given lollies.

And the question that sceptics regularly ask is: 'Haven't we always had extreme weather – floods, droughts, thunderstorms, heatwaves, cold snaps?' 'Well, yes we have.' But climate change will increase the frequency and severity of these extreme weather events. Unfortunately for us all, this is very unintuitive unless we understand these different effects at play and how climate change actually works. And this certainly doesn't help those already burdened with the many normal human reactions such as confirmation bias and the other reasons for denial. So, when an unseasonal cold snap occurs, the sceptics immediately proclaim 'see, I told you! whatever happened to climate change and global warming?' (the confirmation bias trap).

Just as unfortunately, climate change authorities and even scientists often connect weather events with climate change. I assume this is done because scientific facts such as climate change are not easily communicated or understood – they are seeking simplistic examples and 'evidence' that the general public can easily understand and relate to such as the weather. This means both sides are now falling into the same trap. But in their defence, the authorities are damned if they do and damned if they don't. If, in the event of a severe weather event they are asked whether the particular event can be linked to climate change, what are they to say? If they say that 'the event cannot necessarily be directly linked to climate change because climate change is a long-term phenomenon', the sceptics will immediately pounce and declare that

'see, while they keep saying climate change will cause severe weather

events and yet when a severe event occurs, they can't associate it with climate change? So, where is the cause and effect?'

On the other hand, if the authorities declare that:

'yes, the floods last week can be directly linked to climate change.'

The sceptics can just as easily and quite rightly claim that:

'the experts are stretching the truth because surely a single event such as the floods last week does not prove a long-term trend which they say is what climate change will be causing'.

Dealing with a wicked, diabolical problem

Two words, wicked and diabolical, have consistently been associated with climate change as a global planning and implementation challenge.

The term wicked was originally used in social planning to describe a problem that is difficult or impossible to solve because of incomplete, contradictory and changing requirements that are often difficult to recognise. That certainly fits the climate change challenge.

Rittel and Webber's[30] formulation of 'wicked problems' in social policy planning specified 10 characteristics:

1. *There is no definitive formulation of a wicked problem.* This means it is so complex, it's difficult to explain what the problem is. Climate change is complex because apart from the cognitive difficulties associated with scepticism, it crosses economic, social, technical, political and financial boundaries. If there are solutions to climate change, they will undoubtedly require solutions involving all these aspects of our lives.

2. *Wicked problems have no stopping rule.* In other words, there's no end to the solution. This also fits climate change. It will continue. there will be no end to the actions needed to address it.

3. *Solutions to wicked problems are not true or false but good or bad.* We could say climate change is true but that would be simplifying it because there has always been climate change. As for its being good or bad, again, climate change will be bad in many respects but if you're a tomato farmer in Iceland, it will be good.

4. *There is no immediate and no ultimate test of a solution to a wicked problem.* The solution to climate change is extremely complex, evolving and long term. Part of its complexity will be the continuous change needed to address it.

5. *Every solution to a wicked problem is a 'one-shot operation' because there is no opportunity to learn by trial and error, every attempt counts significantly.* This isn't necessarily correct for climate change because 'trial and error' solutions will be inevitable.

6. *Wicked problems do not have an enumerable (or an exhaustively describable) set of potential solutions, nor is there a well-described set of permissible operations that may be incorporated into the plan.* This is certainly true for climate change. Because of its multifaceted impacts, the required actions will need changes in many areas of our economy, energy technologies, politics.

7. *Every wicked problem is essentially unique.* Climate change is a unique problem and one we have not been able to face in our current time. The way we deal with it today is completely different to the way humans dealt with it thousands of years ago.

8. *Every wicked problem may be considered a symptom of another problem.* It can be argued that climate change is a symptom of many other historically recent problems including consumerism, globalisation, technological developments, economic models under which humans have existed since the industrial revolution.

9. *The existence of a discrepancy representing a wicked problem can be explained in numerous ways.* The choice of explanation determines the nature of the problem's resolution. Climate change and its complexity and multifaceted solutions are difficult to explain in a simplistic way.

10. *The social planner has no right to be wrong (i.e., planners are liable for the consequences of the actions they generate).* This is specifically for social planning but still applies to climate change in some way due to the difficulty in measuring and solving specific aspects of climate change.

The above definitions apply more or less, to climate change, mainly because most people don't fully understand the extent of the problem and, even if they do, they don't fully understand how to deal with it. Even if the problem is understood, most people don't want to accept what needs to be done. It's simply too big a problem and the solutions are too difficult and complex to implement and accept. The changes we need to make are enormous and we may not be able to make some of them. This applies particularly to leaders and policymakers, mainly because the many solutions are hard to devise and politically difficult to 'sell'.

First, the problem is enormous and possibly one of the greatest challenges for humanity in the modern era. It is enormous in its extent and impact, and in the vast changes the world has to make to avert predicted catastrophic impacts. Even the solutions that may slow down climate change are vast. Fundamentally, if we are to properly address climate change, we need to revise and change the way we live. The biggest and most difficult change will be in the way our economic model predicates consumption. If we don't consume goods and services, the economy falls apart, resulting in considerable social unrest. And unless all our goods and services contain little or no carbon, this consumption will mean more greenhouse gases.

Secondly, it is diabolical because it will require globally binding agreement and actions – and fast. Until there is globally binding and agreed actions, and major changes in the way we live on the planet, nothing of significance will happen to avert climate change. The changes needed will require extensive reshaping of the way we use energy and consume resources. Not only do we need to change our energy use to low carbon but to use less energy, fewer resources and less of everything. This is because everything we use and consume has some energy content. This will severely affect global economies. If we do this slowly to minimise the social impact, it will be too little too late. And until a strong and binding global agreement on a greenhouse gas emission abatement plan is in place and acted upon, some people and nations will wait for others to move first.

Thirdly, the changes to be made in the world's largely disparate economies will be completely different. So a global agreement on how to make these changes will pose considerable difficulties. On the one hand, we have highly developed and industrialised economies in the West and at the other extreme, countries that barely have enough food and hardly any energy sources. At one extreme we have US and Australian citizens consuming energy at many times the global average, and at the other end, people in developing countries who don't even have electricity. The industrialised world has to become decarbonised, while the nations playing catch-up will need to invest in more expensive renewables.

A big stumbling block in negotiations in the United Nations for a global agreement to reduce greenhouse gas emissions has centred on one single but huge issue: the vast inequality in the world's wealth and resources.

How can the UN ask developing countries to curb their development so that it doesn't increase the unacceptable emissions generated by developed countries? Philosophically, the poorest nations have as much right to develop and become as affluent (and yes, as polluting) as the richest nations. To give you an indication, some of the highest emission countries emit (per person) more than 40 times the greenhouse gases than the lower emitting countries.

Take India for example. Although it is fast developing, considerable parts and population of India don't even have electricity. Here we are, suggesting that we should cut down our use of electricity and use of fossil fuels while huge populations still cook with bits of wood collected from nearby bushland. So, basically, we have a major global problem which we don't fully know how to deal with. It's a wicked and diabolical problem.

To add to all this, climate change is subtle, slow and the solutions will take a long time. In an era of political and economic short-termism and quick fixes, this presents a major obstacle to politicians and policymakers. It can be argued that one of the biggest hurdles to mobilising the global community to act on climate change is short-termism. Quite simply, the enormous task of correcting our life on this planet to address a long-term problem will not be achievable with short-term economics and politics. Almost all aspects of our lives are controlled by short-term thinking. Every corporation in the free democratic world presents its profits and losses on a yearly basis. Every non-profit organisation does the same. Most democratic governments have a three- to four-year term. So all decisions and policies are prepared with a one- to four-year horizon.

Fear of change

Change is arguably one of the most stressful factors of life. Consciously and unconsciously, we get stressed at times of change and mostly avoid it as much as we can. We are creatures of habit and we find comfort in normality and stability in every aspect of our lives. Even positive changes are stressful, like getting a new (better paid) job, moving to a new house, travelling to new places, or getting married.

Sometimes change involves a significant loss and our brains hate loss. When we invest ourselves emotionally in anything, it is harder to change because we don't want to lose the time and effort we have exerted. As a result, we have a hard time letting go of a project we know will fail. We also struggle to end doomed relationships because we're terrible at accepting the whole thing was for nothing. In reality, time isn't wasted because it is part of our learning process, but our brains like to see the entire time as a loss rather than just a part of the inevitable conclusion or part of life's rich tapestry of experience and learning. Studies[31] have shown that losses are twice as powerful, psychologically, as gains. We therefore often prefer to avoid loss than make a gain. So we put much more effort into avoiding loss than investing in potential gains. Ironically, it is change that leads us to a better place, a better relationship, a better job and a better life.

When it comes to climate change, the perception is that to avert the outcomes and reduce the huge impacts, we all need to change our lives considerably and possibly sacrifice our standard of living (loss), at least in the short term. Associated with such perceptions of change and sacrifice is the knowledge that responding to climate change will not only require considerable change in our lifestyle but also significant economic costs. Whenever a proposal is put forward to reduce our reliance on fossil fuels, or replace current technology with more energy-efficient systems, a cost is put on such actions. In many circumstances, new energy-efficient and low-carbon technology are more costly than existing systems. The installation of roof solar panels is a good example – expensive to install but the house owner reaps the benefit later in lower energy bills. So it becomes a matter of short-term financial cost for long-term gain.

So, here, there's another major issue for our decision-making processes and the risks of long term future uncertainty. Because there is always uncertainty in future events, any decision we make for the long term is a difficult one. Consider a simple decision whether to install solar panels on the roof of a house. What if electricity costs change? What if the government provides subsidies in future? What if solar panels become more efficient? What if we move house in the near future? Therefore, even though the costs and benefits for installing solar panels may be clearly favourable, because of our natural tendency to put a higher emphasis on loss and avoid long term gains, such a decision becomes very difficult for most people.

This example of solar panel installation could be considered a simplistic analogy for the philosophical dilemma of the global community. In many ways, the climate change debate revolves around short-term costs (loss) against the long-term gains for future generations. Are we, those currently alive today, willing to change our ways and make a few sacrifices for the protection of future generations?

Guilt (and Blame)

We've seen that political alignment is a big factor in denial or acceptance of climate change science. There is also evidence[32] that there is an age gap in this.

A disproportionate proportion of sceptics globally is in the 50-plus age group. There is also a further statistical peak in white males within this demographic. The following age groupings are instructive:

18–29 years old
- 60% say the Earth is getting warmer because of human activity
- 17% say the Earth is getting warmer but mainly because of natural patterns
- 20% say there is no solid evidence the Earth is getting warmer

30–49 years old
- 55% say the Earth is getting warmer because of human activity
- 22% say the Earth is getting warmer but mainly because of natural patterns
- 21% say there is no solid evidence the Earth is getting warmer

50–64 years old
- 48% say the Earth is getting warmer because of human activity
- 22% say the Earth is getting warmer but mainly because of natural patterns
- 28% say there is no solid evidence the Earth is getting warmer

65+ years old
- 31% say the Earth is getting warmer because of human activity
- 34% say the Earth is getting warmer but mainly because of natural patterns
- 33% say there is no solid evidence the Earth is getting warmer.

The discrepancies between the age groups are considerable but more important is the significance of the older age groups in policymaking and in governments and corporations.

And then there are other factors such as colour and ethnicity which also play a significant part in acceptance or denial. For example, only 44% of 'Whites', 56% of 'Blacks' and 70% of 'Hispanics' say that the earth is getting warmer because of human activity. Furthermore, there is a gap in the sexes where 26% of men compared with 24% of women think there's no evidence of global warming!

A study[33] by the Pew Research Center gives slightly different results but the main picture is the same. So what is it about white, over 50 men and climate change? Let's call this the Group of White Men Over Fifty (GOWMOF) and its subgroup of those in denial (GOWMOFID). While the discrepancy between the age groups may not be as stark as say political or ideological leanings, the importance of the GOWMOFID is that they represent a disproportionately high level of political and corporate power.

The GOWMOF has a special place in the social, cultural, economic and political make-up of Western society. It is also part of the demographic of baby boomers

In Western economies boomers are widely associated with privilege, as many grew up in a time of widespread government subsidies in post-war housing and education, and increasing economic opportunity and affluence. As a group, Boomers were the wealthiest, most active, and most physically fit generation up to the era in which they arrived, and were amongst the first to grow up genuinely expecting the world to improve with time. They were also the generation that due to unprecedented economic growth and opportunity were able to get educated, work hard and receive high levels of income. They could therefore reap the benefits of consumer goods, leisure, food, real estate, retirement programs. It is therefore no surprise that the consumerism for this generation has been regularly criticized as excessive.

It's almost an exclusive club and its members are pretty proud of being in it. Apart from a feeling of belonging, another factor likely to be involved is guilt.

I discussed a few aspects of this group above but an additional observation I've made of this GOWMOF's views on climate change can be summarised as follows: In Western economies, this group has dominated (and still dominates) government policy and corporate management and had the biggest impact on the western developed economies over the past century or so.

So, here we have a global group of people who has dominated corporate and government policy and decisions since the industrial revolution. It is certainly this group that has also been responsible for the new era of commercialism, consumption, unprecedented development and technological advances beyond the imagination of past centuries. So it's perhaps no coincidence that it is mainly this group that has the highest percentage of climate change sceptics.

Not only did they make the political and economic decisions that led to the emissions of greenhouse gases that are now blamed for climate change, but this group has been the cause of these emissions. They bought the big cars, the big houses, travelled the world, used the air conditioners, enjoyed the comfort and safety of a highly energy-intensive lifestyle.

So for the GOWMOF to accept the science is the equivalent of its saying

'We stuffed up. We made the wrong decisions. We thought we were clever. We were so focused on the short-term gains; we didn't factor in the long-term costs and effects.'

or

'we can simply deny that climate change is happening and deny that we are the main cause of it.'

I must emphasise that I'm not suggesting the GOWMOF was the only group that made the policy and corporate decisions that caused greenhouse gas emissions. We could also say the GOWMOF provided for the incredible and fantastic advances in technology and economic growth and consequent improvements in the quality of life. When it comes to economic development, technological advances and improvements in the quality of life, the GOWMOF has been responsible for many things of which to be proud.

One of the many ways some of us overcome the pain of guilt is to pass it on to others. Instead of taking responsibility for the increase in greenhouse gas emissions, we blame successive governments for their inaction, the consumption-based economy, the lack of green choices, oil companies.

In 2010, a major environmental catastrophe occurred in the Gulf of Mexico, the Deepwater Horizon oil spill. A mobile offshore drilling rig operated by BP exploded and sank. A sea-floor oil gusher flowed for 87 days until it was capped. Eleven people died and it was the worst accidental marine oil spill in the history of the petroleum industry. It is estimated that 4.9 million barrels of oil were discharged. After several failed efforts to contain the flow, the well was declared sealed in September 2010. But in early 2012 there were indications that the well was still leaking.

One thing is certain, the environmental damage and the considerable financial impact on many businesses will continue for many years, perhaps decades. In September 2010, I wrote an opinion piece for the ABC website Unleashed. It said:

In London, the BP directors are debating whether to suspend the next round of dividend payments, following the Gulf of Mexico oil spill, whilst the US President is making his fourth visit to the affected region before meeting the company directors on Wednesday. What does this mean for consumers and investors? Let's consider it this way.

While BP's actions (or inactions) dealing with the design, construction and operation of the oil rig are yet to be fully explained and analysed, if negligence is proven, BP faces severe fines – at a cost calculated per barrel. So far, the amount of oil spilled is estimated at between one and two million barrels. Possibly the greatest ever environmental disaster.

Now, ask yourself how we, as consumers and investors, have contributed to increasing the risk of such disasters? What signals do we give to corporations to take these risks? We do this in many ways, and these include:

- *Demanding high returns on our investments and within a short term – it's about money without consideration of the consequences.*

- *Demanding an endless supply of the commodities such as petroleum fuels – fuels which are more and more difficult to access – hence the need to drill thousands of metres below the ocean.*

- *Demanding and expecting these endless supplies of fuel to be cheap so that we can continue to use more of them. We scream, complain and blame governments and everyone else we can point a finger at, when oil prices rise.*

I liken the above global attitude to the adage of wanting our cake (oil), eating it (consuming fuel), wanting more of it (endless supply) – without getting fat (without any environmental consequences).

BP's liabilities have sky-rocketed and their costs to clean up, to pay the pending fines and to compensate those affected, not to mention lost revenue, will be enormous – and their investors will suffer. But should investors in BP be the only ones who pay?

One thing is absolutely certain, and we must face it – resources such as oil will become increasingly scarce; this will increase the costs and the environmental risks of finding more.

Unless we face this reality and accept some personal responsibility, we must accept the increasing risks of environmental disasters with the consequences of them impacting our health and quality of life.

The greatest change that may arise from this oil spill is that individuals may consider how sustainable their investments and lifestyle choices really are. Just look at superannuation in Australia, which is estimated at $1.1 trillion of funds under management and set to rise to $3.2 trillion by 2035.

If we all invested in sustainable options, imagine the wake-up call to the corporate world. We may see the end to the race for oil; the technical development and exploration of more renewable and viable energy sources and more energy efficient processes.

This was one of the objectives of the proposed Emissions Trading Scheme, but it seems we had our eyes on the cake. Perhaps we'd all just prefer to feed our sweet teeth, and get fat?

My point is that we can't simply blame fossil fuel companies for taking risks and exploring for and finding more fossil fuels until we consumers take action to reduce and ultimately eliminate our demand for fossil fuels. So guilt and blame are two sides of the same coin. We either accept at least some responsibility for climate change and act, or we simply pass the guilt to others through blame. The latest recipient of the 'guilt pass' is China. Other than its unprecedented economic development, the main reason China is emitting so much greenhouse gas is that it is now manufacturing many of the products that used to be made by the

Western economies. China has become the factory of the world. So how much blame can we pass on to China or any one nation?

Ideology and politics

I have kept this topic until now, because I wanted to cover the others before getting to the **big one.**

Is there a link between our ideology and our attitudes towards issues such as abortion, healthcare, social welfare? The answer is a resounding yes, and this has been well established by a number of studies.

We had a glimpse of the significance of ideology on the responses to climate change in 'I don't like the solution above. When it comes to climate change, there is considerable evidence that ideology, not evidence, drives our acceptance of the science, and whether and how we should respond to the predicted risks. The reality is that politics, or ideology, is by far the most influential factor in accepting or denying climate change science.

So, what is it about ideology that influences our understanding and acceptance of climate change science? Generally, our ideology is how we see the world we live in, and how we believe it should be in terms of social, economic, religious and political ideals.

There is also something about climate change and greenhouse gas emissions that differentiates them from, say, air pollution. Air pollution is clearly demonstrated to be directly harmful to our health and it makes a lot of sense, after all, we breath the air. At worst, it is toxic and at best it has unpleasant or long-term undesirable effects. It is mostly generated and emitted as a result of industrial activity. By contrast, the main greenhouse gas, carbon dioxide, although it is a pollutant and an asphyxiant in high concentrations, is also part of what's termed the carbon cycle[34]. It occurs naturally in the atmosphere and is generated by volcanoes, waste decomposition, humans and animals. It is necessary for the ecosystem. But greenhouse gases include water vapour so, already, we have a definitional problem. Greenhouse gases include natural as well as anthropogenic emissions.

Please bear with me while I go through this issue as it is as interesting as it is complicated.

Liberal (left wing) versus conservative (right wing) views of the world
Broadly speaking, right-wing conservative ideology advocates for the right to individual determination and expression (sometimes including gun ownership), cutting taxes, lowering government spending and balancing the budget. Right-wing politics generally opposes socialism and social democracy and maintains that some forms of social stratification or inequality are inevitable, natural, normal or desirable, typically defending this position based on natural law, economics or tradition. After all, inequality is everywhere in nature. And when it comes to economics, even what is termed 'trickle down economics' advocates and assumes that benefits for the wealthy ultimately benefit everyone else.

Again, speaking broadly, left-wing ideology supports egalitarianism, often in opposition to social hierarchy and inequality. It typically involves concern for those who are disadvantaged compared with others and a belief that inequalities need to be reduced or abolished. It upholds the principle that human development flourishes when individuals engage in co-operative, mutually respectful relations and excessive differences in status, power and wealth are eliminated.

So what does all this have to do with the acceptance, or otherwise, of climate change science? Well, as it turns out, quite a lot.

A Study of studies
Several studies have suggested a strong link between ideology and the acceptance of climate change science. I'll mention a few.

The connection between our ideology and topics such as abortion, market based economic policy, social welfare and free speech are all well established. They, after all, deal with how we see the world and how we feel our society should be ideally structured and how it should function. However, the correlation between a person's ideology or political leanings and his/her acceptance of scientific facts is an interesting phenomenon. It is almost like saying that accepting the threat of cancer or heart disease is linked to one's political views, but not quiet.

A major meta-analysis[35]

A major study recently published in *Nature* found that belief in climate change science is neatly divided along political and ideological lines. The researchers, from the University of Queensland[36], examined 27 variables by synthesising 25 polls and 171 academic studies from 56 nations.

The research analysed in the study has covered a wide range of disciplines, including psychology, communication, sociology, political science, agriculture, climate science, and media studies. It summarised the relationship between climate change belief and 7 demographic variables, 13 psychological variables that, according to theory, should be what has led to climate change belief, and 7 variables widely considered to be downstream consequences of climate change belief.

The study extracted broad themes. As such, it provides a comprehensive overview of who endorses or opposes the reality of climate change and the main reasons they do so. First, many intuitively understood variables (such as education, sex, subjective knowledge and direct experience of extreme weather events) were overshadowed by values, ideologies, world views and political orientation. Second, accepting climate change science has only a small to moderate effect on the extent to which people are willing to act in climate-friendly ways. Implications for converting sceptics to the climate change cause, and for converting believers' intentions into action are also discussed in the study.

Some other findings of the study include:

- The biggest demographic factor in climate change belief is political affiliation. People who intend to vote for more left-wing political parties are more likely to believe in climate change than those who see themselves as conservative.
- The connection between political alignments and acceptance of climate change is about double the size of any other demographic

variable. The tendency for right wing people to express more scepticism than left wing people has long been identified within the US. This has contributed to a growing ideological gap between sceptics and non-sceptics.

- Relatively small effects were found for the other variables: age, education, income, race and sex. People with stronger beliefs in climate change were younger were more educated, had higher incomes, and were more likely to be non-white and female.

- Although a 'conservative white male' profile has emerged of climate change sceptics in the US, the analysis of polls across multiple nations suggests that the 'conservative' part of that equation would seem to be more diagnostic than the 'white male' part.

- Despite the common perception that climate sceptics are mainly older white men, there is not a strong correlation with any other demographic but ideology. Education, gender, general knowledge and experience of extreme weather had little effect.

The lead author of the paper, psychology professor Matthew Hornsey, told the Washington Post: 'People higher in scepticism are more likely to be old, white and male – but the effects are so tiny you have to squint to see them. What really popped was people's ideologies, political values, worldviews.'

George Lakoff, a University of California, Berkeley-based cognitive scientist told ThinkProgress that the ideological difference is:

'nothing surprising for anyone who actually does cognitive science. It has to do with the way in which conservative versus progressive values are understood unconsciously.'

Some people:

'seem to believe that it's all simply getting people the facts, and they will reason to the right conclusion. It doesn't work that way. Brains

don't work that way.'

Climate change activists see the world differently to sceptics. The difference is stark. According to the paper, there is a strongly positive link between knowledge and belief in anthropogenic climate change, but only for Democrats and independents. So the more knowledgeable independents and (left wing) Democrats reported themselves to be, the more likely they were to accept the science of climate change. For (right wing) Republicans, the level of reported knowledge didn't matter.

But while facts alone are not going to win the day and drive climate action, facts still matter to some degree. While general facts and studies don't seem to change beliefs, real-life observable threats such as droughts, storm surges and wildfires can be motivators. Even for people who aren't convinced that higher greenhouse gas emissions are fuelling extreme weather, a sceptic might think climate action is 'a bet I'm willing to take'.

If anything, this report points out how hard it is to get people of either ideology to act. Even for people who accept the science on climate change, there was only a slim effect on willingness to take climate-friendly action. In other words, people who believe that we are making the Earth uninhabitable for large sections of humanity aren't much more likely than outright sceptics to actually do anything about it.

American ideology

A review published in the *European Journal of American Studies* by Jean-Daniel Collomb[37] concludes that although the origins and motives of the US climate change denial movement are complex, they illustrate the strong ideological forces that have been shaping Republican politics in the past few decades, mainly because climate change science:

'significantly damages the soundness of the ideological pro-market position which the American conservative movement has been embracing since the Reagan era and the end of the Cold War'.

The reason behind this is that although the argument that human activities contribute to climate change is not necessarily contrary to a market economy, it does question the validity of the free market as the ultimate solution to all social, economic and environmental problems. In fact, a number of prominent economists, such as Sir Nicholas Stern, have argued that climate change is a failing of the free market economy because it hasn't put an economic value or cost on the damage to the environment caused by human activity.

In a way, recent economic models have put short-term economic gain ahead of long-term loss. We have simply ignored the environmental downside of our consumption patterns and resource development, including the establishment of mines, urban development and the destruction of forests. It's like eating cheaper fast (junk) food now and ignoring the long-term costs of treating major illness, or at least deferring them to the future. Ironically, we put a cost to almost everything else, even life itself. So, for example, while we consider the costs of keeping an elderly or terminally ill person alive, we are failing to put a cost for damaging the environment and the huge future costs of repair.

Because the most influential element of scepticism by far is ideology, climate change discussion has become aligned with other social and political issues such as abortion, gun control, healthcare and migration.

This divides our society into two camps. There is really no middle ground. Even those who claim to be undecided are as discussed earlier, in effect, sceptics because while they make up their minds – if they ever do – they know nothing will be done about climate change.

When people say to me 'I'm still not sure about it', it is clear to me that they still are in denial and are yet to be convinced. And until they are convinced, if ever, then they are in denial. It's akin to saying to oneself: 'despite all the tests and my doctors' advice, I'm really not sure that I have cancer'.

Spreading the word

Social change and fundamental shifts in public policy involve considerable effort, require robust conversations in the public arena, debate in and outside politics, and take time to implement. While leadership at all levels of government are important, in a democracy the public needs to be involved and must be taken along the path to change. Public support of policy change is paramount and needs to be much better than a 50/50, or even a 60/40, split between support and opposition. The problem with climate change is that this needs to happen quickly because the longer it takes, the more that must be done and the more difficult it will become.

To obtain public support and participation will require considerable debate and delivery of information. But for the debate to be meaningful and productive, the information must be accurate and credible. And the media can and should play a part in providing information and encouraging debate. But is the media in a position to know or judge what is accurate or credible?

The media's role in spreading the word and engaging proper and robust debate on complex issues such as climate change isn't easy and has proved very difficult. It appears to have become even more difficult with shorter turnaround of news, comment and opinion.

We need to come to terms with what needs to change, what we must do to make these changes and how to manage their outcomes. Problem arise when the commentary and opinions are disparate and contradictory. Another problem is which advice, opinion and comment should we give more weight to. As we've seen above, policymakers and the public may give more importance to their preferred political and social leanings. While we need to listen to all opinions and commentary, not all advice is equal and sometimes, some advice is more equal than others.

In the past few decades, many of the numerous social changes in the Western world have been highly controversial and divisive. They include emotionally charged issues such as abortion, marriage equality, terrorism, immigration and refugee intake. The climate change debate has gone on for many years with seemingly little progress, or at least progress that is too slow for some.

We've said before that responding to climate change requires fundamental shifts in the way we live and, the way we obtain and use energy. We know that the changes will require considerable global, regional and national efforts and that this won't be easy – for many reasons. One of the largest reasons being the difficulty in attaining global agreements and targets for energy usage, which require agreements at the regional and national levels. The biggest changes will be needed in developed countries such as Australia.

So how do we begin? One of our greatest challenges is communication to encourage robust and balanced discussion and debate, galvanise public and governmental opinion and then make big policy changes on numerous aspects of our lifestyle. We need to do this in a post-truth world where 'fake news' can travel much faster than informed and credible news and where going viral is more important than credible and rigorous discourse. And credible journalism can suffer when it has to compete with the speed of social networks and as a result has less time to check the validity and credibility of 'real' news.

The media can, knowingly or unwittingly, fuel uncertainty and confusion on climate change science and feed the negative aspects of human responses to the complexity of climate change. Due to the complexities of getting the messages across the general public, the media can muddy the waters and, at worst, mislead and misinform. But how can it avoid doing this?

Its most important impact has been in its perceived need to provide the widest possible spectrum of views and opinions. In trying to be inclusive and broad in its coverage, the media has given a voice to less than credible, extreme, and at times, totally discredited points of view on climate change. This problem is not isolated to the way the media deals with climate change; this relativist approach occurs in other areas of discussion too.

Is this good journalism? Some may argue it isn't. But how can it be done better? Is it reasonable for us to expect journalists and editors to write and publish only credible and reasonable articles?

I'm all for democracy and free speech but a topic such as climate change requires public support and courageous political will to implement measures that will undoubtedly be unpopular with some members of the community. Such support and will are not attainable while there is confusion about the science and misunderstanding about the impacts and the required actions. In August 2007 Kevin Rudd, who was then the Australian opposition leader, delivered a speech to the National Climate Change Summit. He said 'climate change is the great moral challenge of our generation'. While this was heartfelt and for many people true, the public was still agnostic about the subject. So Rudd wasn't able to galvanise support, mainly because apart from the presence of a strong sceptics club within politics, the public didn't adequately understand the issue or appreciate the costs and benefits to Australia.

The media has also had another agenda. We know that good news isn't anywhere near as interesting as bad news. Good news just doesn't sell. In addition, conflicting and diverse opinions are more newsworthy than generally accepted views and facts. If you look at a normal distribution curve, otherwise known as the bell curve, the middle isn't anywhere near as interesting as the outer edges of the curve. For example, we accept the science that links respiratory disease with smoking. If someone provides additional data to link smoking with respiratory disease, that's news but not very compelling. However, if a scientist (credible or not) has carried out some research (published or not) that indicates possible health benefits of smoking, that's real news. That is, even if the data shows that such possible benefits may only apply in rare circumstances and to very small part of the population. While such news may be seen to be worthy of publication, it has significant impact for those looking for an excuse to continue their addiction to continue smoking.

How many times have I seen a TV interview with an internationally respected and published expert on climate change, and a high-profile personality (such as a lord from the UK) arguing about whether climate change is occurring. The personality's knowledge is a fraction of the expert's and, given enough time, the personality's argument would be destroyed by the expert. But the program allows only a few minutes for the discussion. The personality brings up one or two points of contention in a brief statement, which might be a misquote or an extracted figure from a document that is either not published or not peer reviewed. The expert needs a lot more time to completely demolish the other's argument. So often the arguments remain without a clear resolution. The viewer is usually no wiser and any doubts in their mind remain.

I'm not sure whether we expect the media to have a role in informing and educating the public on complex and important issues such as climate change. And I'm not saying that all media do not present balanced and credible stories.

But if the media were genuinely interested in having such a role, on television, it would dedicate an hour or more to a panel of internationally respected experts on climate change to answer fundamental questions raised by sceptics. It could have three sceptics facing three experts. In the print media, prominent experts would provide comprehensive answers to the types of questions sceptics ask. Or they could simply reprint the many well-documented website explanations by numerous authorities and academic institutions. On radio, prominent experts could explain the science in simple terms and answer questions that listeners pose.

While we have established that being informed doesn't necessarily help people 'believe' the science, misinformation and confusion certainly don't.

In fulfilling the need to expose all views and opinions, the media has, in effect, given equal voice or, at least disproportional voice, to marginal climate change observers or scientists. Thousands of scientists and experts have spent decades researching, gathering data, modelling and publishing thousands of peer-reviewed papers on countless aspects of climate change. By comparison, a small number of people, some of them scientists, have cherry-picked aspects or assumptions of some published papers or reports to question. So, here we have those that have scientifically proven climate change is taking place and has been caused by human activity; and those that challenge these findings. Despite the huge imbalance of the number, scientific rigor, and credibility of between the two groups, they get almost equal share – or at least disproportional share in the news. In fact, to some extent, the smaller, more radical, more rebellious group gets more of a say, much like the smoking example. The rebels get a disproportionate say because it's deemed more newsworthy.

Polarising the readers may not be a conscious or deliberate strategy for the media but it certainly sells news.

Another recent phenomenon and yet another contributor to the outsiders' voice issue is the feedback and comment opportunity of the broader opinion base. Almost every popular media outlet invites the public to provide feedback, commentary and opinion online. Not to mention on social networks, which are increasingly viewed as media outlets. If such opinion were provided across the entire population base and were representative of the whole population, it may be a reasonable means of gathering views on contentious topics. But experts are highly unlikely to be wasting their time writing comments on these sites. They have better things to do. On the other hand, the radicals provide plenty of material.

In addition, even if we could obtain views representative of the entire population and gather representative set of opinions on a topic such as climate change, would we want that? On this highly complex topic, does the opinion of the average person matter or should the widely spread opinions of the population have a bearing on decisions and policymaking? Is this a requirement of democracy?

Actually, I contend that in my generous statement above regarding a representation of the entire population, even with an even spread of the population's views, many would not consider this to be a reasonable voice of the science of climate change. In such a complex field, we need experts, to tell us the facts rather than having the entire population giving its disparate and relatively uneducated opinions. If we were after the latest science or facts about breast or prostate cancer, we wouldn't be relying on the general population or the media to give their opinion or views! Public policy affects our economy and the way we live and must surely be based on proper scientific evidence and scholarship.

We'll find a way

As indicated above, one of the groups some belong to in the climate change debate is one that says:

> 'Yes, we are contributing to climate change but humanity is capable of solving this. We don't have the answers now, so why not leave it to future generations (and technologies) to solve?'.

There is some justification for this thinking. We live in such rapidly changing times. Just look at what humans have achieved since the industrial revolution. Our technological development has been exponential in its impact and efficacy. In the past 100 years – a mere blip in the timeline of human existence on this planet – we have invented the computer, can send people to the moon and bring them back, fly large, metallic vehicles from one end of the world to the other in less than 24 hours, send documents and images from one end of the world to the other in a fraction of a second, perform extraordinary surgical procedures and replace somebody's heart with another's. Ironically, this development has been one of the factors that has led to the build-up of greenhouse gases in our atmosphere. Still it is possible that new technologies will find some of the solutions to reduce the greenhouse gases in the atmosphere. Already, we are finding ways to harness renewable energy and this is an area of very rapid advances. But are they quick enough?

Medical science has led to our longevity increasing rapidly. In the Bronze and Iron Ages, life expectancy was about 26 years; in 2010 it was 67.2. The combination of high infant mortality and deaths in young adulthood from accidents, epidemics, plagues, wars and childbirth, particularly before modern medicine was widely available, significantly lowered life expectancy. Ironically, it's once we survive the early years that we begin to consume more resources and burn more fossil fuels – when we begin to buy motor vehicles, use air travel, build houses and reproduce more consumers.

Not only are we increasing the length of our lives by treating almost every disease until old age, but preventing childhood mortality. It wasn't too long ago that a significant number of babies died before or soon after birth. But mothers also died at birth. Antibiotics have saved hundreds of millions of lives. And yes, the growing human population is a contributing factor in the consumption of resources and the emission of greenhouse gases.

But humans are adaptive. We seem to find a solution to major problems, one way or another. The human race has survived for many millenniums and now that we have all this technology, surely we can find a way out of this seemingly huge problem. And, as this appears to be a slow burner, a problem that will creep up gradually not suddenly, surely we will find solutions and adapt as we have always done.

This kind of discussion ends up being a matter for intergenerational equity. Is it fair for the people alive today to leave the problem of climate change to future generations to solve? It's fair to say that this generation alone didn't create this problem. It can be argued that several generations (since the industrial revolution) have created this problem. But it can also be argued that this is the generation that uncovered the problem and the one that can do something about it. Is it willing to do that?

We need to ask a broader question: have we lost touch with nature? Have we now become so arrogant and detached from the environment that we allow it to deteriorate without any conscious thought about our long-term survival? Not so long ago, humans were hunters and gatherers and relied entirely on the environment and our interaction with nature to survive. Now, in a strange way – and to a large extent – the environment depends on us.

Apart from medical science and many other human endeavours, our engineering innovations and efficiencies in mass production and manufacturing combined with economic development have meant that more people around the world can afford to consume more. This has resulted in a huge increase in the ownership of motor vehicles, and use of electrical appliances and air travel, all of which contribute to greenhouse gas emissions.

So, what are the solutions that may somehow avert the predicted impacts of climate change? Can we rely on them?

Certainly, solar power and the efficiency with which we can harness this endless source of energy will continue and a major breakthrough is possible. There are also considerable opportunities in harnessing wave or tidal power. But the biggest opportunity in its sheer potential for baseload power generation is geothermal energy. While all this is taking place, we will no doubt continue to debate the benefits and risks of nuclear energy.

In his book *Earth Masters*[38], Clive Hamilton details several investigations into the feasibility of 'geoengineering', large-scale interventions to counter global warming or at least offset its impacts. Such concepts would include the removal of carbon dioxide from the atmosphere, a type of clean-up. Another involves reducing the amount of solar radiation reaching the planet, a masking of the problem. It is possible that we will find ways to deal with the problems of greenhouse gases and the effects of climate change. But the question is whether we will find solutions in time, because the longer we leave the problem, the worse it will get, and the more difficult it will become to solve. An additional question is whether such innovative solutions, even if they were realised, would be less costly than the solutions we already have available.

Part C

Where
to
Now?

Introduction

In this Part I provide just a few ideas on the direction we need to be heading if we are to properly address climate change. The technology is already available and improving at great speed to decarbonise the world. The economic and social consequences are more complex to deal with. Nevertheless, there are countless ways the world can respond and each geographical and economic region, each country, and individuals will have their own means and solutions. What follows are a few of them.

A combination of factors

Humans are complicated beings. We are driven by all sorts of biases, fears, conceptions and misconceptions, ideology, personal beliefs, religion, ego – you name it. When we are confronted by complex issues that affect not only ourselves but those around us, we revert to our basic instincts and drivers for social, economic and spiritual comfort and security.

So the challenges about communicating on climate change are due not only to the complexity of the science but are made even more diabolical by the complexities of human cognitive tendencies. And these cognitive issues are often combined in any individual or group of people. They simply pile up on top of each other. If you look carefully at the reasons given above, like me I'm sure you'll find yourself and at least some of your tendencies among them.

Now that we are researching and investigating the many cognitive and behavioural factors in the communication of climate change science, we may eventually find a bright light at the end of the tunnel. I'm certain we will find a means to communicate the details of climate science and obtain agreement for positive action to ensure that future generations don't have to deal with the increasingly difficult tasks of adapting to it.

What we've learnt so far

What we have learnt so far is that climate change is a difficult issue and a diabolical problem for the world to solve quickly. There is plenty of evidence that we are more than capable of eventually decarbonising the world, delinking economic growth from greenhouse gas emissions, and adapting to the significant climate change coming our way. Whether we do all this in time is questionable. What is not open to debate is that the longer it takes us to act, the harder it will be for future generations.

What we have also learnt is that communicating the right messages to the world community to obtain global consensus and agreement has been extremely difficult and at times ineffective. Due to a number of cognitive responses, there have been numerous barriers to the messages that needed to be understood and agreed upon by world leaders.

There will necessarily be the need for short-term pain for long-term gain to secure direct and urgent action. Right now, the world is lengthening the short-term gain to reduce the economic shocks and subsequent social effects. Unfortunately, stretching out the short-term pain will probably increase the pain for future generations. It's like ignoring a decaying tooth.

Renewable energy technology, particularly solar power, is rapidly becoming more efficient. On top of this, the cost of the power generated is falling substantially so that in some instances, solar power is more cost effective than many other forms of energy supply, particularly in isolated communities.

The concept of carbon neutrality has caught on. It refers to organisations, processes or products that have zero net emissions – a zero carbon footprint. Usually, this is achieved by using renewable energy for the entire process or production, or offsetting net greenhouse gas emissions by buying offsets to 'neutralise' the emissions. The most common offsets are those that result from renewable energy generation, forest or land management or other means of sequestering or capturing and storing greenhouse gases.

There is a number of accreditation systems to ensure such offsets are real and additional to what would have normally taken place. For example, if a company invests capital to install a wind turbine to generate electricity in an area that would have obtained its power from a coal-fired electricity generator, that company would be given carbon credits equivalent to the wind energy generated and would be able to sell its carbon credits to another organisation as an offset.

Offsets have been particularly successful in the airline industry where passengers can pay extra for their airfare to offset the greenhouse gas emissions of their flight. Ironically, in effect this is putting a price on carbon emissions, a price many people objected to when introduced as a pollution tax, yet happily paid by numerous air passengers.

Energy efficiency in homes, commercial buildings and industrial complexes has had some success but has a long way to go. There is considerable evidence of huge opportunities for the industrialised world to become much more energy efficient and to design more sustainable cities, buildings and manufacturing processes.

We know that information alone will not sway people's views and galvanise them to action. Part of the challenge is overcoming the antipathy towards politics in general, globalisation, authorities such as the UN, and the general community feeling that:

The 'system' is feeding itself and betraying us.

The main problems to solve if we are to properly address global warming are:

1. ensuring that global decisions and actions are seen to be for the good of everyone or at least for most, and not perceived to be only for the bureaucratic elite

2. obtaining global community support for more ambitious and more urgent action to decarbonise developed as well as developing economies

3. securing global leadership to set and deliver ambitious and binding targets for decarbonising

4. setting up a more aggressive and influential global organisation than the UN Framework Convention on Climate Change (UNFCCC) to achieve the required agreements, set the ambitious targets and deliver the greenhouse gas abatement measures required

5. decoupling economic stability and the consumption of resources and greenhouse gas emissions; addressing climate change must be viewed as an economic prerogative rather than a threat to the economy.

A 2015 international report[39] detailing how individual countries, including Australia, could transition to a low-carbon economy found limiting global warming to 2 degrees Celsuis by 2050 was still achievable. The report suggests that all deep decarbonisation pathways incorporate 'three pillars' of energy system transformation:

1. energy efficiency and conservation
2. decarbonising electricity and fuels
3. switching end uses to low-carbon supplies.

The report emphasises that deep decarbonisation cannot be achieved if any of the pillars is absent or implemented at insufficient scale.

Energy efficiency and conservation: The report found that energy efficiency reduced the energy intensity of gross domestic product by an average of 65%, with nearly all countries making their economies two to four times more energy efficient in 2050 than in 2010. This was accomplished through measures such as improving vehicle fuel economy, better building design and construction materials, more efficient appliances and industrial processes and machinery, and conservation measures such as urban design to encourage walking and bicycling.

Decarbonising electricity: In all deep decarbonisation pathways (DDPs), electricity becomes nearly carbon free by 2050, with average emissions per kWh reduced by a factor of 15 below the 2010 value. This was accomplished by progressively replacing most uncontrolled fossil fuel-based electricity generation with varying mixes of renewable energy, such as wind, solar, geothermal and hydropower, nuclear power and fossil-fuel generation with carbon capture and storage. Liquid and gas fuel supplies were decarbonised using biomass fuels with low embedded carbon emissions and synthetic fuels such as hydrogen produced from decarbonised electricity.

Low carbon supplies: The dominant trend in final energy consumption is to replace coal and petroleum with electricity and lower carbon fuels, including a coal to natural gas shift in some DDPs. Much of the direct combustion of fossil fuels in end-use equipment such as vehicles, hot water heaters and industrial boilers is replaced by decarbonised electricity, which more than doubles the share of electricity in final energy consumption in 2050 to more than 40%.

Where do we need to be?

Above, I gave a very brief summary of where we are now and promised to give some key suggestions of where we need to be.

If we had a great big magic wand, we would wave it and convert all our energy supply to renewables and all our problems would be solved. We could use as much as energy as we liked and other than the risks of running out of resources, such as minerals and the potential impacts on the environment, global warming could be averted or at least minimised. But we don't have a great big magic wand, we don't even have a small one.

The solutions and possible actions to reduce global greenhouse emissions are far too numerous to give a complete list here. And each region, country and state will need to respond in their own way. What follows is aimed at providing examples of measures and a broad set of possible actions. It's certainly not a 'to do' list. On an extremely simplistic level, what needs to be done, and quickly, is as follows.

Global actions and initiatives

The world needs a better institutional system to deal with issues such as global warming and climate change. The present system is too clunky, too bureaucratic, too slow and is not able to act appropriately and quickly with this major global problem. If a huge meteor were heading our way and was likely to hit Earth within a few years, how would the world respond?

1. Global leadership must change its dialogue to ensure that actions to address climate change are not seen as part of a huge bureaucratic rhetoric and yet another part of globalisation.

2. The economic necessity of action must be fully estimated, explained and emphasised as much as the environmental and social impacts. Although this has in part been done, the message isn't getting through.

3. Coal-fired power stations must progressively close and be decommissioned. All planned coal-fired power stations must be abandoned. But as we still need energy to drive everything, this obviously cannot be done without the commissioning and operation of new renewable energy plants to replace those that are being closed. So renewable energy supply needs to take place in parallel with the decommissioning of fossil-fuel stations. This is a huge task that will take time but it needs to happen as quickly as the world is able to do it without social and economic upheaval. But some short- to long-term economic costs may be inevitable.

4. Consideration must be given to the establishment of nuclear power stations, particularly if this becomes necessary to replace the decommissioned coal-fired stations. This is a highly divisive and controversial issue and requires careful planning, particularly relating to the storage of waste material. Unlike many types of renewable energy, nuclear plants take a long time to get going. So this cannot be done without the commissioning and operation of new renewable energy plants to provide for the increasing demand for electricity.

5. Discontinue the subsidisation and support of fossil fuels. In a report released by the Overseas Development Institute and by Oil Change International in November 2015, it was revealed that G20 governments were providing US$444 billion a year in subsidies for the production of fossil fuels. This is bad economics – many argue that propping up any industry through subsidies is not good economics – and the support should instead be given to encourage the development of renewable energy and energy-efficient technologies.

6. The economies of many nations rely heavily on fossil fuels. The transition to a low-carbon world will not be easy. To minimise social and economic dislocations and hardship, it will have to be done gradually and will require careful, considered and planned replacement of fossil fuels with renewables and, at the same time, the reduction of the energy intensity of all goods and services. The problem is that to do it gradually will risk considerable long-term hardship caused by climate change.

7. Continue (more quickly than is being done now) the development of renewable energy plants to keep up with global population growth and economic development, particularly in developing countries. This will necessitate continuous and focused research into and the development of new technologies to efficiently convert

solar and wind power to electricity as well as harness geothermal energy and the storage of energy, particularly for solar power.

8. Continue research and development of energy efficiency of buildings, processes and manufacturing as well as transport, lighting and air conditioning. There needs to be a dramatic reduction in the energy intensity of all our activities.

9. Significantly reduce consumption of energy, resources and materials by decoupling consumption and economic development. At present, the term economic development is synonymous with the production of goods and services and their consumption. While this will remain of primary importance for social and economic development and stability, there is a need to reduce overconsumption and wastage of resources. We need to remember that all materials we consume have embedded energy.

10. Instil the notion of environmental and climate change risks in all economic and business decisions. This will necessitate the decoupling of environmental and economic decision making. Corporations and governments must include climate change and environmental risks in all financial decisions. At present accounting standards don't include the environmental risks of investments and operations. The long-term effects of operations and the cost of reparation and remediation must be included in all financial accounts and reports.

11. A number of carbon-neutral initiatives provide incentives for organisations to aim for and achieve carbon neutrality. These include the Gold Standard, the Verified Carbon Standard (VCS) and, in Australia, the National Carbon Offset Standard (NCOS). Carbon neutrality is usually achieved through a combination

of measures to reduce energy usage, use green power and renewable energy.

Corporations around the world voluntarily become partly or completely carbon neutral for various reasons, including cost reduction, marketing, stakeholder expectations and social responsibility.

In 2012 Microsoft introduced a carbon price by imposing an internal carbon fee that holds business units accountable for their emissions. This cost becomes a line item in their annual budgets. It also set an incremental fee on emissions from the company's global operations such as data centres, offices, laboratories, manufacturing and air travel. The aim is to be carbon neutral, so the annual fee is based on the estimated cost of internal efficiency, renewable energy, carbon offset and e-waste recycling. Some of the funds collected are used to place renewable energy systems in developing countries.

Google's net carbon footprint in 2013 was zero. The company achieved this through buying green power as well as carbon offsets. In 2014, its long-term agreements for renewable power produced 1,056,433 MWh. These contracts covered 28% of the company's total electricity consumption. It also obtains green power from the grid and on-site renewables, making the total share of renewables more than 37%. Google has made agreements to fund nearly $2.5 billion in renewable energy projects. The total capacity of its renewable energy project investments is more than 3.7 GW. These projects are expected to generate more than 11 billion kWh of electricity annually, far more than they consume.

The above suggestions are not meant to be easy fixes and as summarised earlier, much good progress is being made. But perhaps many of them will need to be hastened.

Regional initiatives

Regionally binding agreements are needed throughout the world to reduce greenhouse gas emissions through achievable targets. The agreements that have been implemented through the European Union and the European Commission are good examples and include carbon trading policies and support mechanisms. There are many more examples of actions being taken by regional and other groups.

Building on the results of the 2015 Paris climate summit, in July 2016 G20 energy ministers reconvened in China to discuss further strategies for creating a low-emission future.

Another example is the treaty signed by 14 countries in July 2016 at the annual leaders summit of the Pacific Islands Development Forum. It agreed to consider a Pacific climate treaty that would bind signatories to targets for renewable energy and ban new coal mines or the expansion of existing ones.

There are a number of US regional action such as the Western Climate Initiatives (WCI) by five western US states to develop and implement market-based programs to reduce greenhouse gas emissions. Another is the Regional Greenhouse Gas Initiative (RGGI) which is the first mandatory market based program in the US to reduce greenhouse gas emissions.

National initiatives

1. Each country needs to revise its commitment to decarbonise. Such a revision must take into account the anticipated long-term predictions of the impact of climate change. It won't be easy to do this and there is likely to be many economic adjustments and teething problems. Major changes to any fundamental way we live and use energy is bound to be fraught with obstacles. But we need to find a way through them.

2. Decarbonisation must happen through policies and incentives to install more sources of renewable energy and disincentives to use

fossil-fuel sources. Putting a long-term cost on greenhouse gas emissions and climate change must be part of such policymaking. This will encourage investment in low-carbon energy and technologies and, at the same time, provide direct subsidies and incentives in research and development.

3. Economists agree that an effective way to encourage renewable energy and discourage the continued use of carbon-intensive and inefficient systems is the development and implementation of a price on carbon emissions. The idea being to impose a cost on carbon emitting processes and encourage renewable energy and energy efficient processes. The money raised from such a policy can be put towards research in renewable and energy-efficient technologies.

4. Studies need to be undertaken to update the costs of inaction and of decarbonising. Serious discussion is needed at the highest levels of government and academia about the levels at which the present older generation is willing to pass on the risks and costs of climate change to future generations. Climate change is a giant inter-generational equity issue and the views and concerns of young people must be considered during these discussions.

Individual actions

Individually, we are unable to control global agreements and actions. But we can exercise our democratic rights to influence, encourage and support local, state and federal governments to take action to reduce greenhouse gases.

Direct and personal measures

There are some things we can all do to reduce our direct contributions to greenhouse gas emissions. The most obvious is through our use of fossil fuels.

1. Use of petroleum for transport

    ☼ Next time you buy a car, choose a smaller and more fuel-
        efficient one. Also remember that the weight of a car affects
        its fuel usage, mainly because it takes more energy to move a
        heavy weight around. All else being equal, a heavy car is less
        fuel efficient than a light one.
    ☼ If you can afford it, next time you buy a car, choose an electric
        or hybrid model.
    ☼ Use the car as little as you can. Walk instead of driving, ride
        a bike or take public transport when practical or convenient.
        Riding a bike and walking also have obvious health benefits
        and save money. A number of car insurance companies offer
        discounts for low usage.
    ☼ Use your car efficiently by driving calmly, accelerating gently.
        The way you drive your car has a significant effect on your fuel
        usage not to mention wear and tear.
    ☼ Minimise the use of air conditioning in your car. Air conditioning
        systems use energy.
    ☼ If possible, organise or take part in car-pooling to travel to and
        from work. This can also save you money.
    ☼ If possible and convenient, get rid of your car altogether and
        join a car-sharing system. Depending on your circumstances,
        these options can be healthy, convenient and save money.

2. Heating and cooling your house

    ☼ This presents a significant proportion of our energy costs and
        contributes considerably to greenhouse gas emissions. Using
        natural gas and oil for heating emits greenhouse gases and
        isn't the most efficient way to keep warm. Consider installing
        reverse cycle air conditioning. Keep windows, shutters and
        blinds closed and avoid droughts to keep the heat in in winter
        and out in summer.

☼ Don't over-heat or over-cool your house. A degree or two cooler in winter and a degree or two degrees warmer in summer will make a significant difference in your costs of heating and cooling.

☼ Shutting off areas of the house not being used. When heating or cooling the house, this will also save a lot of energy and money. Considerable heat is lost in winter and gained in summer through windows and roofs. Install insulation in your roof, put blinds and curtains on your windows and block gaps to stop leakage of energy and keep blinds and curtains closed on hot and cold days. At the very least, all roof spaces should be insulated. The payback for such investment is often within months. In some cases, it may even be cost effective to install double-glazed windows.

3. Hot water

☼ Hot water is another significant proportion of our total household energy use (and cost). When your existing hot water system dies, choose an efficient one.

☼ Minimise the unnecessary use of hot water. Use warm or cold water when they would do just as well.

☼ If your hot water system has a storage tank, you may be able to set the temperature lower and this will significantly reduce your costs.

4. Appliances

☼ Choose energy and water-efficient appliances. Major electrical appliances that use a lot of energy include the refrigerator, dishwasher, washing machine, space heater, water heater and TV. Water efficiency in a dishwasher and washing machine also affects energy usage when heating and hot water are involved.

☼ Switch off appliances and lights when they're not needed. This sounds like an obvious thing to do but many people still keep appliances and lights on unnecessarily –wasting energy and money.

5. Renewable energy

☼ Install solar photovoltaic panels on your roof. You could reduce your greenhouse gas emissions to zero and become part of the global renewable energy network. You may even be able to return some renewable energy to the electricity grid.

☼ There is a payback period for such an investment but ultimately, you'll be using free energy and minimising your carbon footprint.

Indirect measures

Individually, we contribute significantly to indirect emissions of greenhouse gases. For example, when we switch on an appliance (using electricity from a coal-fired power station), the electricity we use results in greenhouse gas emissions at the point where the power is being generated.

Another form of indirect emissions we are all responsible for is the waste we generate and the goods we buy. Everything we buy and use has an energy content and this usually includes:

- mining of metals and minerals
- processing of raw materials
- transport of materials to the manufacturing plant
- manufacturing processes
- transport of manufactured goods to the market, and
- packaging.

When we dispose of these goods, particularly organic material such as food, they usually end up in landfill where they generate methane and carbon dioxide – both are greenhouse gases that mostly end up in the atmosphere. This is known as the life-cycle greenhouse-gas emissions of goods. As an exercise, consider the life cycle and cradle to grave energy content of a sheet of paper or a packet of chips. You may be surprised by the number of steps needed to produce a simple product, each one of which involves the use of energy.

Here are a few ideas on how you can reduce your indirect emissions. Most of these will also save you money.

1. Buy green power. This not only reduces your own carbon footprint but also encourages the renewable energy market and associated technological developments. The higher the demand for green or renewable power, the more investment will take place in low carbon power technologies.

2. Buy low carbon or carbon-neutral products and services. Organisations that have achieved carbon neutrality freely advertise such achievements but many other products and services are low carbon and it is worth investigating the carbon content of your service providers and the products you buy.

3. Buy locally produced goods. This reduces transport related-energy and emissions and supports the local economy.

4. Buy locally grown food products. Such food products are usually fresher as they haven't had to travel far and generally haven't been picked or harvested early.

5. Buy produce that is in season. Out-of-season produce is either imported and travels a long way, or is grown in local greenhouses which usually use energy.

6. Buy food more carefully – whenever possible, in small amounts. This will reduce wastage and also save you money.

7. If possible, compost your organic food waste. This will stop it going to landfill and generating methane and carbon dioxide. It will also give you a wonderful soil conditioner to use on your garden.

8. Reuse and recycle materials as much as you can. Glass, metals, plastics and other packaging have a lot of embedded energy. Reusing and recycling not only conserve materials but also reduce carbon emissions.

Jumping the hurdles

Economic justification

Much has been said about the cost of decarbonising global economies. There is no doubt that investing in renewable energy and decommissioning coal-fired power stations will involve a lot of money. On top of that, there are costs associated with changing our transport systems and investing in more efficient and accessible public transport.

The Stern Review, among many others, identified the costs of inaction.

The review assessed a wide range of evidence on the impacts of climate change and on the economic costs, and has used a number of different techniques to assess costs and risks. From all of these perspectives, the evidence gathered by the review leads to a simple conclusion: the benefits of strong and early action far outweigh the economic costs of not acting.

Using the results from formal economic models, the review estimates that if we don't act, the overall costs and risks of climate change will be equivalent to losing at least 5% of global GDP each year, now and forever. If a wider range of risks and impacts is taken into account, the estimates of damage could rise to 20% of GDP or more. In contrast, the costs of action – reducing greenhouse gas emissions to avoid the worst impacts of climate change – can be limited to around 1% of global GDP each year.

It can also be argued that not enough has been said about the health costs of *not* decarbonising the world. When we decarbonise, we will eliminate a huge amount of air pollution. Fossil fuels emit not only carbon dioxide but many other harmful pollutants into the air. We all breathe this pollution into our lungs every time we take a breath. The air pollution from burning fossil fuels has a huge cost to our health. A recent study[40] by Harvard researchers calculated net benefits of US$38 billion a year under a plan proposed by the US Environmental Protection Agency to regulate carbon pollution from existing power plants. And that's just power plants. That does not include many other highly polluting industries and operations. And that's just in the US.

Put simply, decarbonising the world will have health benefits that can be quantified in monetary terms. It will clean the air of pollution, reduce many diseases and mitigate climate change. It will cost money in the short and medium term, but will reap many benefits to future generations.

We know climate change will also have a huge cost associated with damage to vital infrastructure and this is also being estimated. Extreme heat, floods, storms and sea level rise will all have significant impact on major infrastructure around the world. This is in addition to the economic and social impacts on agriculture, human health, food and water supply. Calculating the global cost of adaptation remains a complex problem. In a recent publication, the World Bank41 selected just 7 developing countries and estimated:

the price tag between 2010 and 2050 for adapting to an approximately 2 degrees Celsius warmer world by 2050 will be in the range of $70 billion to $100 billion a year. Our country studies suggest that costs could be even higher, once cross-sectoral impacts are taken into account.

Public and people power (Brexit, Trump)

Some of the lessons of recent political and social phenomena such as Brexit and the Trump ascendancy have already been discussed. It is evident from recent world events and, in particular from the changing political environment that there appears to be a move against globalisation and authoritarianism, and for reclaiming national identity and local control.

Part of the backlash against climate change for political leaders is that even if they want to implement more urgent actions to abate greenhouse gas emissions, it is difficult to gain a mandate. So for the past decade or so, the strategy has been 'easy does it'. It can be argued that most of the initiatives for energy efficiency and renewable energy programs have been initiated by corporations. The rest have come slowly, almost by stealth, so that the apparent or perceived economic impacts are minimised.

Many would argue that getting a little done is better than nothing done. In an article in *The Conversation*,[42] in September 2016 Clive Hamilton suggests:

> *There are the pragmatists willing to compromise to get at least something, and then there are the idealists who stick to their principles and end up with nothing … It's all very noble, the pundits have been writing, to stick to what climate science demands, but in the real world of hard politics what we need now is a way through the political impasse.*

So, what is needed for a seismic shift to take place is a global wake-up call and political will by all major emitters. This can happen only if, and when, there is full public support in these countries for more positive and urgent action.

We have learnt that for such a major shift to take place, information or education are not the key. Some fundamental shifts that could help the world deal with the challenges of addressing climate change are:

Money talks, but it can also walk

As the population lives longer, superannuation becomes more important for our retirement and the quality of our latter years. Total assets of the world's largest 300 pension funds grew by more than 3% in 2014 (compared to about 6% in 2013 and 10% in 2012) to reach a new high of more than US$15 trillion. So globally, superannuation funds manage a huge amount of money which is invested in corporate shares, property and other financial products.

Wouldn't it be great if much of this money was invested in renewables, energy efficiency and energy storage technologies? It may also at least partly overcome the generational equity issue due to the funds belonging mostly to the generation that mostly created the climate change problem.

Lengthen the term

Short termism is a huge barrier to addressing issues such as climate change. Climate change is a long-term problem and the myriad of potential solutions are mostly long term. Take the idea of putting a price on carbon emissions as suggested by many prominent economists. The idea isn't just about penalising carbon emitters, although that's how it will provide a disincentive to emit greenhouse gases and provide incentives to decarbonise. Even such incentives will not bear fruit for many years. But much more than that, the idea is to provide incentives for technological development of energy efficiency and low carbon processes. On top of that, it is meant to put some of the collected money from polluters into scientific research and into providing subsidies for low carbon commercialisation of new technologies. All of this will take time.

Governments and corporations necessarily have a short-term focus on their decisions, policies and operations. But as voters and consumers, we can all put more pressure to widen the focus, and lengthen the payback periods and planning horizons.

Support for a price on carbon

Most reputable economists – even conservative ones – agree that the concept of financial incentives that support market-based mechanisms are an effective way to reduce greenhouse gas emissions. The concept puts a price on greenhouse gas emissions – a polluter pays idea. Yes, some would consider it a penalty on the emission of carbon. But another way of looking at it is that it puts a price on the environment, something Nicholas Stern has said we have avoided doing. It can also be seen to be saving the enormous future costs of dealing with the effects of climate change.

Doing our bit

There's little to be gained by simply blaming politicians, the fossil fuel industry and the United Nations. We all need to take personal responsibility to improve our energy efficiency and reducing our carbon emissions. We can also educate ourselves and become more informed about climate change, energy efficiency and carbon emissions. Then we can pass on our knowledge to those around us and pressure our leaders to get on board.

We can also play a part in interest groups such as the Australian Youth Climate Coalition (AYCC) and the Youth Climate Movement (YouNGO) or International Youth Climate Movement (IYCM) and the proper communication and discussion of climate change science. Based on an article published in the World Metrological Organization[43] I suggest we can all do this:

1. When you see a post or article that challenges or disputes climate change science, point the author to the many peer reviewed publications, credible web sites and books on climate science. There are far too many to mention more than just a few here:

   - United National Framework Convention on Climate Change (UNFCCC)
   - Intergovernmental Panel on Climate Change (IPCC)
   - USA Environment Protection Agency (USEPA)
   - The World Bank

- Climate Council
- Australian Bureau of Meteorology
- CSIRO
- The Climate Institute
- European Climate Change Program
- Australian Academy of Science
- American National Academies
- National Aeronautics and Space Administration (NASA)
- The Royal Society

2. All the common misconceptions, myths and doubts about climate change can be clearly articulated and there are several websites including some of the above that do this well including:
   - https://www.skepticalscience.com/argument.php
   - http://grist.org/series/skeptics/

3. As we have discussed, linking extreme weather with climate change is not simple.

   Begin the discussion with what is known rather than starting with uncertainties, and what we cannot say. For example:

   'We know that in a warming world, we experience more frequent and severe heat waves. And we see that trend clearly in the data. This event is part of that trend.'

   Then discuss any studies relating to the specific extreme weather event being discussed, such as those that quantify the altered chances of the event. For example:

   'Global warming made this heat wave at least four times more likely to occur, or increased the odds of this event by 400%.'

4. Communicate clearly and simply the mechanisms behind the changes brought on by warming. For example:

'A warmer atmosphere holds more moisture, leading to heavier rainfall.'

5. Use metaphors to help explain how human-induced warming changes the odds of extreme weather events. For example:

'heat-trapping gases act like steroids in the climate system, increasing the odds of extreme heat, heavy downpours and other types of extreme events. We're now experiencing the weather on steroids.'

Emphasise that even though extreme events do occur naturally, many are now happening more frequently and more intensely. Similarly, global warming 'is stacking the deck' in favour of such outcomes and 'we know climate change is happening now, and is human-caused, even if we can't be certain that it is a direct cause of a particular event.'

Another metaphor is this:

'imagine a thin invisible blanket around the Earth which stops the heat from escaping. The more the greenhouse gases, the thicker the blanket.'

6. Remember that many words mean entirely different things to scientists than they do to the public. For example, scientists often use the word *uncertainty* to discuss the envelope of future climate scenarios, or the range of model results for a particular finding. But to the general public, *uncertainty is* usually interpreted as *we just don't know.*

   Referring to *a range* is better than calling it *uncertainty*. Similarly, scientists may describe a finding as being *low confidence* to do with data or model issues. But this does not mean there is no observed trend or no projected change as the public might assume from such language.

7. Try to avoid language that can lead to despair and hence inaction. Rather than calling further increases in extreme weather *inevitable*, we can discuss the choice we face between a future with more climate change and larger increases in extreme weather, and one with less.

Scandinavian culture, a good example

In his book *Affluenza*, Oliver James, a psychologist, compares modern-day influences that encourage affluence with an infectious virus (influenza).

This 'affluenza', caused by our propensity to needlessly consume, manifests as distress and general unhappiness. He suggests we should look towards the Scandinavian countries where relationships and the public good are put first. Scandinavian countries are, overall, well-organised, efficient, and aesthetically appealing countries in which to live – not to mention quite environmentally sustainable. In contrast, we are all working very hard to pay for things we want but don't really need. We are looking for *more* instead of being happy with *enough*.

The main problem here is that we are in a spiral that will be difficult to escape. Consumption of goods and services makes global economies go around. The spiral starts with the need to consume goods and services. To pay for these things, we need to work long hours and hard. Because we spend so much time working, we don't have much time for things such as cooking, exercising and family. So, we spend more time in gyms and buy ready-made meals or go out to eat. We outsource a lot of other chores and activities, such as cleaning our house and looking after our children. This stresses us out and we might end up becoming sick because we don't spend enough time on the things that matter. So then we need to work a little harder to pay for more things including taking holidays to 'recharge our batteries'. But ideally, we must gradually escape of this spiral. Wouldn't it be nice if we needed less money and worked less to spend more time with our loved ones and things we really wanted to do?

James looks at how some cultures such as Denmark's manage to create and live a better, balanced life. Their values are more attuned to gender equality, glamorous rather than sexy, natural rather than artificial, to looking after the elderly – and the environment. In 2015, Denmark had the highest top personal income tax rate among the 34 countries in the Organisation for Economic Co-operation and Development. The personal income tax rate, which is now 55.8%, averaged 61.12% from 1995 until 2015. It reached a high of 65.9% in 1997 and a low of 55.4% in 2010.

You might say; what does all that have to do with addressing climate change? I contend that the way we live, our values and life choices, our long-term view of social justice and intergenerational equity, all have a lot to do with how we live our lives on this planet.

If you've ever visited your local landfill waste disposal site – and you should – you'll see the amount of waste your community generates. It's really not a pretty sight. But it's not only unnecessarily filling holes in the ground that has environmental impacts. It's more to do with just how much we consume and then discard. It's also about the consumption of goods we don't really need.

Clean Up Australia as a good example

One of the best examples of the power of the global community is the Clean Up movement. The idea began with an Australian sailor being disturbed by the amount of rubbish in our waterways and oceans. Now the event takes place yearly in about 150 countries with more than 35 million people taking part.

The success of this movement relates to the general population's reaction to untidiness and litter. People going for a swim or a sail want to do it in pristine waters. The main advantage of this problem is its sheer visual impact. We can all see the rubbish and we certainly are affected by it every time we go for a swim or sail on a boat, yacht or ship.

This campaign has shown that the average person can do something about the problem. And the result of our action is immediately visible. One reaches over and takes a floating piece of plastic and puts it in a rubbish bin. The problem is removed and the result is there for all to see, immediately. The problem is identified, action is taken, and the result is there to see. Furthermore, we know the cause of the problem and we can see its effects much more clearly than those of climate change. We know that someone threw away a plastic bottle or it perhaps fell off onto the water, and we can see its visual impact. We have all also see images of fish and birdlife tangled in bits of plastic or swallowed pieces of rubbish.

The single most important thing we can learn from the Clean Up movement is that it is possible to get positive and direct action at the community level and then gain support from international organisations and governments. It's a great example of 'bottom up' strategy and action.

Part D

Concluding Remarks

I had several motives for writing this book. First, I became fascinated and intrigued by the number of climate change sceptics I encountered, as I worked in climate change consulting over the past 10 years or more and much longer as an environmental risk consultant. I asked myself why these people were unable to accept the science. Why were they so passionate about their scepticism? Without wanting to find people to blame, I also wanted to find if there were inadequacies in the communication between the climate scientific community and the general public.

Secondly, as the years went by, I watched the highs and lows of the climate change debate, and world politics. I became increasingly concerned about the inability of the world to overcome the complexity of addressing climate change. It seemed – and still seems too difficult for global agreement and effective action.

So I began investigating the reasons for climate change denial. I wanted to know how and why seemingly intelligent, educated and often well-informed people who, despite the ever-mounting evidence, still find it difficult to accept the science. What is going on, I asked myself. Climate change may be difficult for some to comprehend intuitively but why can't some people simply accept the well-established science?

I already knew that climate change was never going to be an easy path for the global community to deal with. The necessary actions are enormous and require fundamental changes to the way we live on this planet. But when I began to research the psychology of scepticism, I began to understand the cognitive barriers for accepting the science. And this simply compounded the numerous changes and actions required to slow down and mitigate global warming and climate change.

So I also began to wonder how, if at all, we could have or still can, communicate and further galvanise global agreement and action. I asked myself whether there was a lesson we could learn about the failure of the debate?

I have included a summary of the fundamental principles of climate change and its impacts as well as the national and global renewable energy initiatives to give a context to the sections on denial. I have also made brief suggestions on the direction we need to go to decarbonise and some of the ways we can all take action. The solutions are many and complex and I never intended to cover them all in this book. I did this because in my early discussions with friends and colleagues, it was suggested that the readers may want to know what needs to be be done and what they can do.

Ultimately, I might have raised more questions than I have answered. But I hope I have at least shed some light on the cognitive and human aspects of climate change. I also hope that if we better understand the barriers that are apparently at play, we can have a different conversation and carry out further research on how to better galvanise global action. Perhaps we can then apply the lessons learnt to other equally difficult global issues.

Bibliography

Bast, E., Doukas, A., Pickard, S., van der Burg, L., and Whitely, S., *Empty Promises*. Overseas Development Institute. November 2015.

Bloomquist, K.L., with Machila, R., *God, Creation and Climate Change, A resource for reflection and discussion*. On behalf of the Lutheran World Federation, A Communion of Churches, Department for Theology and Studies, Geneva, 2009.

Botzen, W.J.W., van den Bergh, J.C.J.M. Bouwer, L.M., Climate change and increased risk for the insurance sector: a global perspective and an assessment for the Netherlands, *Journal of the International Society for the Prevention and Mitigation of Natural Hazards,* 8 May 2009.

Buonocore, J.J., Lambert, K.F., Burtraw D., Sekar S., Driscoll C.T., *An Analysis of Costs and Health Co-Benefits for a US Power Plant Carbon Standard*. (2016)

Callaghan, P., *Science, Ideology and Climate Change*. The University of Adelaide. School of Psychology. Submitted in fulfilment of the requirement for the degree of doctor of philosophy. September 2014.

Campbell, T.H., and Kay, A.A., Solution Aversion: On the Relation Between Ideology and Motivated Disbelief. *Journal of Personality and Social Psychology*. Vol. 107, No. 5. 2014.

Collomb, J-D., *The Ideology of Climate Change Denial in the United States*. European Journal of American Studies. Vol 9, No.1 2014.

Eder, A.B., Fiedler, K., Hamm-Eder, K., Illusory correlations revisited: The role of pseudocontingencies and working-memory capacity. *The Quarterly Journal of Experimental Psychology*. Volume 64, 2011, Issue 3

Encyclical Letter *Laudato Si of the Holy Father Francis on Care for our Common Home*. May 2015.

George Mason University Center for Climate Change Communication. *Climate Change in the American Christian Mind*. March, 2015

George Mason University Center for Climate Change Communication. *Global Warming, God and the 'End Times'*. 26 July, 2016

Hamilton, C., *Earthmasters: Playing God with the Climate*. Allen & Unwin. 2013.

Hamilton, C., *Earthmasters: The Dawn of the Age of Climate Engineering*. Yale University Press. February, 2013.

Hamilton, C., *Requiem for a Species: Why We Resist the Truth about Climate Change*. April Earthscan. 2010.

Harvey, C., Science confirms it: Denial of climate change is all about the politics. *The Washington Post*, Energy and Environment, February 2016.

Hoffman, A.J., Climate Science as Culture War. *Stanford Social Innovation Review*. Fall 2012.

Hornsey, M.J., Harris, E.A., Bain, P.G., and Fielding, K.S., Meta-analyses of the determinants and outcomes of belief in climate change. *Nature, Climate Change*. February 2016.

Intergovernmental Panel on Climate Change (IPCC). *Fifth Assessment Report (AR5)*. 2014.

International Energy Agency. *Global EV Outlook 2016: Beyond one million electric cars*.

International Energy Agency. *Key Energy Statistics 2015*.

International Renewable Energy Agency (IRENA), *Renewable Energy Statistics*, 2016.

International Renewable Energy Agency (IRENA), *Unlocking Renewable Energy Investment: The Role of Risk Mitigation and Structured Finance*, 2016.

James, O., *Affluenza: How to be Successful and Stay Sane*. Vermillion. 2007

Kahan, D.M., Climate-science Communication and the Measurement Problem. *Advances in Political Psychology*. Vol. 36, Suppl. 1, 2015.

Kahan, D.M., Ordinary Science Intelligence: A Science Comprehension Measure for Use in the Study of Risk Perception and Science Communication, with notes on evolution and climate change. Yale Law School, Yale University, New Haven, CT, USA. *Journal of Risk Research*, 2016.

Kahan, D.M., What is the 'science of science communication'?, *Journal of Science Communication*, Vol. 14, No. 3, 2015.

Kahan, D.M., Peters, E., Wittlen, M., Slovic, P. Larrimore Ouellette, L., Braman, D., and Mandel, G. The polarising impact of science literacy and numeracy on perceived climate change risks. *Nature*, Climate Change. May 2012.

Knight, E., *Why We Argue About Climate Change*. Redback, 2013.

McNeil, B., *Clean Industrial Revolution: Growing Australian Prosperity in a Greenhouse Age*. Allen & Unwin. 2009.

Monmouth University Polling Institute. Public Says Climate Change is Real. 5 January, 2016

Oreskes, N., & Conway, E.M., *Merchants of Doubt*. Bloomsbury Press. 2010

Pew Research Center *How Different Groups Think About Scientific Issues*. 12 February, 2015.

Pew Research Center *More Say There is Solid Evidence of Global Warming*. 15 October, 2015.

REN21 Renewable Energy Network for the 21st Century. *Renewables 2016: Global Status Report*.

Rittel, H.W.J. and Webber, M.M., Dilemmas in a General Theory of Planning. Policy Sciences. 4, 1973.

Steffen, W., *Super-Charged Storms in Australia: The Influence of Climate Change*. Climate Council. November, 2016

Stern, Sir N., *Stern Review: The Economics of Climate Change*. October 2006.

Stoknes, P.E., *What We Think When We Try Not to Think About Global Warming*: toward a new psychology of climate action. Chelsea Green Publishing. 2015.

Sustainable Development Solutions Network (SDSN) and the Institute for Sustainable Development and International Relations (IDDRI), *The Deep Decarbonisation Pathways Project 2015 Synthesis Report*. September 2015.

World Bank, *Economics of Adaptation to Climate Change*: Synthesis Report. 2010.

World Economic Forum, *The Global Risks Report 2016*. 11th Edition. 2016.

Appendix

Summary of global agreements and actions to date

Once the alarm bells started ringing loudly in the mid-1970s, the clear and severe predictions of global impacts of climate change required world agreements and actions. So a number of international bodies were created to clarify the latest science, identify the extent of the problems and articulate a long-term path to address them.

UNEP and the IPCC

In 1972, the United Nations Environment Program (UNEP) was established as the UN's designated entity for addressing global environmental issues. Its mandate is:

'To coordinate the development of environmental policy consensus by keeping the global environment under review and bringing emerging issues to the attention of governments and the international community for action.'

Climate change soon became one of the more urgent issues for UNEP and, in 1988, along with the World Meteorological Organisation (WMO), UNEP established the Intergovernmental Panel on Climate Change (IPCC) to prepare assessments on all aspects of climate change and its impacts with a view to formulating realistic response strategies.

Since then, the IPCC has published five assessment reports. These are composed of the full scientific and technical assessment of climate change, generally in three volumes, one for each of the working groups of the IPCC, plus a Synthesis Report. Each of the working group volumes has individual chapters, an optional Technical Summary and a Summary for Policymakers. The Synthesis Report integrates material contained in the Assessment and Special Reports and is written in a non-technical style suitable for policymakers. It addresses a broad range of relevant but policy-neutral questions. It comprises a longer report and the Summary for Policymakers.

These reports are prepared and reviewed by thousands of the smartest and best qualified scientists, academics and researchers and provide a complete picture of the status of greenhouse gas emissions and their impact. These reports are fundamental to our understanding of climate change, human contributions to it, the challenges and the importance of international co-operation to tackle the predicted consequences.

The UNFCCC

The United Nations Convention on Climate Change (UNFCCC) became effective on 21 March, 1994. It is one of three conventions adopted at the Rio earth summit in 1992 and its ultimate aim is preventing 'dangerous' human interference in the climate system. It has near-universal membership. The 197 countries that have ratified the convention are called parties to the convention.

The genesis of the UNFCCC was remarkable for its time. This is because in 1994, there was considerably less scientific evidence about global warming than there is now. The UNFCCC borrowed an important line from one of the most successful multilateral environmental treaties (the Montreal Protocol of 1987). It bound member states to act in the interests of human safety even in the face of scientific uncertainty. I suspect this courageous step was taken because the world didn't realise that while the first step was easy, the follow-up would be diabolically difficult.

The ultimate objective of the convention is to stabilise greenhouse gas concentrations *'at a level that would prevent dangerous anthropogenic (human-induced) interference with the climate system.'*

It states:

'Such a level should be achieved within a time frame sufficient to allow ecosystems to adapt naturally to climate change, to ensure that food production is not threatened, and to enable economic development to proceed in a sustainable manner.'

And how are we to know what is dangerous anthropogenic interference? The best answer is through the IPCC assessments. So, there's a link between all these conventions, panels and programs.

Under the convention:

- industrialised nations agree to support climate change activities in developing countries by providing financial support for action on climate change, above and beyond any financial assistance they already provide to these countries
- a system of grants and loans has been set up
- industrialised countries agree to share technology with less-advanced nations.

These objectives were in effect made operational through the Kyoto Protocol adopted on 11 December, 1997. Due to a complex ratification process, it took until 16 February, 2005 for it to become effective.

It commits industrialised countries to stabilise greenhouse gas emissions based on the principles of the convention. The Kyoto Protocol sets binding emission reduction targets for 37 industrialised countries and the European Union in its first commitment period. Overall, these targets add up to an average 5% emissions reduction compared to 1990 levels over the period 2008 to 2012 (the first commitment period). As it turns out, this target although achievable, is highly inadequate.

The Paris Agreement reached in December 2015 is the latest within the framework of the UNFCCC dealing with greenhouse gas emissions mitigation, adaptation and finance starting in 2020. Since 22 April, 2016 180 UNFCCC members have signed the treaty. The agreement establishes a long-term goal of net-zero emissions, a mechanism to review progress and increase targets at regular intervals, and a framework for climate finance. The agreement will take effect only after 55 nations responsible for 55% of greenhouse gas emissions have ratified it, making it binding. China and the US, responsible for about 38% of global emissions ratified the agreement in September 2016. At this date, only 26 parties representing 39% of emissions had ratified the agreement.

Despite the excitement of the Paris Agreement and ratification by the two biggest emitters, it is merely an agreement to reduce global temperature rise to 2C. The long-term signal in the Paris Agreement is for countries to peak their emissions as soon as possible and achieve net-zero emissions by the last half of the century. At the heart of the Paris Agreement are the intended nationally determined contributions (INDCs or NDCs), or national commitments, from 186 nations comprising 90% of the world's greenhouse gas emissions. These commitments are to be updated every five years, which leaves the door open to more aggressive emission reductions in the future. But many scientists consider the wording and commitments far too vague for rapid and aggressive abatement of emissions. As most realise by now, even with this unprecedented commitment, we are on a probable path towards at least a 2.7 degrees Celsius global average temperature above pre-industrial levels. This doesn't match the goal of limiting global warming to well below 2 degrees Celsius, or even to 1.5 degrees Celsius. Indeed, some scientists suggest 1.5 degree Celsius is already 'baked into the system'.

The Kyoto Protocol

The central feature of the Kyoto Protocol is its requirement that countries limit or reduce their greenhouse gas emissions. By setting such targets, emission reductions took on economic value. To help countries meet their targets and encourage the private sector and developing countries to contribute to emission reduction efforts, negotiators of the protocol included three market-based mechanisms – emissions trading, the Clean Development Mechanism (CDM) and Joint Implementation (JI).

Considering the huge challenges, the achievements of the protocol were far higher and wider than the casual observer might have realised. Not least were the transfer of technologies from developed to developing countries, the establishment of complex methods, known as methodologies, that clearly establish the technical means of a baseline (business as usual) and then the estimation of emissions abatement, and the implementation of projects that might never have taken place. These methodologies established what is known as the 'additionality' of the projects. This makes sure that without the protocol's punitive encouragement (through the issuance of credits), the project would not have taken place. After all, if a project were financially and technically viable, it should or would have taken place and would not have needed the intervention or encouragement of the protocol's emission credits. Even as claimed by some critics, perhaps rightly, that some of these projects would have eventually taken place, the protocol nevertheless encouraged their implementation earlier than would have otherwise happened.

Once the protocol was agreed upon and emission reduction by developed countries negotiated, it was a matter of getting projects up and running on the two main mechanisms: CDM and JI. Being on the roster of international experts, I was directly involved with many of the projects and with the organisations that dealt with auditing and verifying the emission reduction and their accreditation. There were numerous challenges in getting this done, not least of which were the methods for measuring baseline emissions and estimating the emissions that were abated through numerous projects. Again, I was directly involved with the establishment of many methodologies.

By far the greatest challenge – and perhaps we should call it a barrier to agreement on real and achievable actions – relate to the huge divergence between the industrialised world and the developing world. The difference between the 'haves' and 'have nots' means that not only are they playing catch-up with economic growth and industrialisation (and that means consuming a lot more energy), they are far less able to adapt to a changing climate. Poor nations simply don't have the economic and technical resources to adapt to climate change.

About 17% of the world's population doesn't even have electricity. The simplified reality is this: the developed world has arguably been the main contributor to greenhouse gas emissions and this has been the main driver in the Kyoto agreement, which basically gave the developed world the responsibility (and targets) to reduce emissions. But since then, developing countries such as China, India and other South-East Asian and South American countries (such as Brazil) are fast developing and becoming major emitters of greenhouse gases. So one of the many difficulties in reaching agreements is how does the world agree on the division of responsibility and the efforts required by the various developed and developing countries? Let's face it, it will take enormous effort and extremely ambitious targets for the abatement of emissions by all countries if major climate change is to be prevented. Is it then fair to limit the development (and emissions) of developing countries as part of this ambitious plan? In a nutshell, don't the developed countries have the right to become as affluent as the developed countries? Yes, but in doing so, unless they develop using only renewable energy, the targets cannot possibly be achieved. The problem is that while they are quickly coming down in costs, renewables are still significantly more expensive than fossil fuel and especially coal.

Those who criticise the activities of the UN don't fully appreciate the difficulties in dealing with global issues involving the diverse and large number of countries. And when it comes to climate change and the difficulties of agreeing and acting on climate change, this is no easy matter.

Despite these positive global initiatives and considerable efforts to consolidate the science on climate – and the efforts to agree on equitable and workable action to reduce emissions– we are still nowhere near the targets needed to reach the levels of greenhouse gas reductions.

Notes

1 Throughout this book, the term climate change will refer to anthropogenic climate change, that is, climate change due to human activity.

2 Cook, J., Nuccitelli, D., Green, S. A., Richardson, M., Winkler, B., Painting, R., Way, R., Jacobs, P., & Skuce, A. Quantifying the Consensus on Anthropogenic Global Warming in the Scientific Literature. Environmental Research Letters, 8. 2013.

3 Monmouth University Poll. 5 January, 2016

4 Caused or produced by humans

5 Throughout this book, I will sometimes use the word rejection, denial or non-acceptance (of the science) to mean the same thing.

6 What I mean by 'linearly' is in a continuous straight line. For example, on a graph, a linear increase of data would go from lower left hand corner to the top right hand corner.

7 Latent and sensible heat are types of energy released or absorbed in the atmosphere. Latent heat is related to changes in phase between liquids, gases, and solids. Sensible heat is related to changes in temperature of a gas or object with no change in phase.

8 See http://www.c2es.org/

9 See https://www.weforum.org/

10 See also phenology, the study of cyclic and seasonal natural phenomena, especially in relation to climate and plant and animal life.

11 International Energy Agency. Global EV Outlook 2016; Beyond One million electric cars.

12 Renewable Energy Statistics, 2015. IRENA

13 Reported by Agora Energiewende, a German clean-energy think tank

14 http://ec.europa.eu/eurostat/statistics-explained/index.php/Renewable_energy_statistics

15 http://www.bloomberg.com/news/articles/2016-06-01/chile-has-so-much-solar-energy-it-s-giving-it-away-for-free

16 Biomass most often refers to plants or plant-based materials that are not used for food or feed. As an energy source, biomass can either be used directly via combustion to produce heat, or indirectly after converting it to various forms of biofuel. Burning biomass emits greenhouse gases. The simplest form of biomass for energy is the burning of wood, which is still the only source of fuel for domestic use in many developing countries. Conversion of biomass to biofuel can be achieved by different methods, which are broadly classified into thermal, chemical and biochemical.

17 Feed-in tariffs are usually paid under government schemes to non-commercial producers of electricity generated by solar photovoltaic (PV) systems using solar panels. They are a way of subsidising and encouraging uptake of renewable energy.

18 https://www.scottishrenewables.com/sectors/renewables-in-numbers/

19 Key Energy Statistics. International Energy Agency. 2015.

20 The Organisation for Economic Co-operation and Development, comprising 35 member countries

21 Klaus Fiedler, Eder AB, Hamm-Eder S Illusory correlations revisited: The role of pseudocontingencies and working-memory capacity. The Quarterly Journal of Experimental Psychology. Volume 64, 2011, Issue 3.

22 Pew Research Center Region and Science. Highly religious Americans are less likely than others to see conflict between faith and science. Cary Funk and Becka A. Alper. 22 October, 2015.

23 George Mason University Center for Climate Change Communication. Climate Change in the American Christian Mind. March 2015.

24 George Mason University Center for Climate Change Communication. Global Warming, God, and the 'End Times'. 26 July, 2016.

25 Laudato si'; Praise be to you – On the Care for Our Common Home.

26 God, Creation and Climate Change. The Lutheran World Federation – A Communion of Churches. Geneva 2009.

27 Solution Aversion: On the Relation Between Ideology and Motivated Disbelief, Journal of Personality and Social Psychology. November 2014.

28 Australians are one of the highest per capita emitters of greenhouse gases.

29 The polarizing impact of science literacy and numeracy on perceived climate change risks. Dan M. Kahan, Ellen Peters, Maggie Wittlen, Paul Slovic, Lisa Larrimore Ouellette, Donald Braman and Gregory Mandel. Nature Climate Change. May 2012.

30 Dilemmas in a General Theory of Planning. Horst W. J. Rittel and Melvin M. Webber. Policy Sciences. 4 1973.

31 Kahneman, D. and Tversky, A. (1992) Advances in prospect theory: Cumulative representation of uncertainty. Journal of Risk and Uncertainty. Kahneman, D. and Tversky, A. (1984). Choices, Values, and Frames. American Psychologist.

32 Pew Research Center How Different Groups Think About Scientific Issues. 12 February, 2015

33 The Pew Research Center More Say There is Solid Evidence of Global Warming. 15 October, 2015.

34 The carbon cycle is the process by which carbon compounds are interconverted in the environment, involving the incorporation of carbon dioxide in living tissue by photosynthesis and its return to the atmosphere through respiration and the decay of dead organisms.

35 A meta-analysis uses a statistical approach to combine the results from multiple studies in an effort to increase power (over individual studies), improve estimates of the size of the effect and/or to resolve uncertainty when reports disagree.

36 Matthew J Hornsey, Emily A. Harris, Paul G. Bain & Kelly S. Fielding.

37 Jean-Daniel Collomb, The Ideology of Climate Change Denial in the United States, European Journal of American studies, Vol 9, No1; 2014.

38 Clive Hamilton. Earthmasters: The Dawn of the Age of Climate Engineering. Yale University Press. February, 2013.

39 The Deep Decarbonisation Pathways Project 2015 Synthesis Report. Published by the Sustainable Development Solutions Network (SDSN) and the Institute for Sustainable Development and International Relations (IDDRI), September 2015.

40 An Analysis of Costs and Health Co-Benefits for a US Power Plant Carbon Standard. Buonocore JJ, Lambert KF, Burtraw D, Sekar S, Driscoll CT (2016).

41 Economics of Adaptation to Climate Change: Synthesis Report. The World Bank (2010).

42 http://theconversation.com/climate-policys-house-of-cards-65623.

43 (Un)Natural Disasters: Communicating Linkages Between Extreme Events and Climate Change. Susan Joy Hassol, Simon Torok, Sophie Lewis and Patrick Luganda. Bulletin, Vol. 65 (2), 2016.

Lightning Source UK Ltd.
Milton Keynes UK
UKHW021814070819
347572UK00017B/348/P